Biodiversity Loss and Ecosystem Collapse: A Global Assessment Through 15 National Case Studies

Azhar ul Haque Sario

Copyright

First Printing, 2025

Azhar.sario@hotmail.co.uk

ORCID: https://orcid.org/0009-0004-8629-830X
LinkedIn: https://www.linkedin.com/in/azharulhaquesario/

Disclaimer: This book is free from AI use. The cover was designed in Canva.

Disclaimer: This book is an independently produced work of analysis and commentary. The author has no affiliation with, is not sponsored by, and is not endorsed by the World Wildlife Fund (WWF), the Intergovernmental Science-Policy Platform on Biodiversity and Ecosystem Services (IPBES), the World Bank, the Convention on Biological Diversity (CBD), or any other organization mentioned herein. The use of organizational names is for identification purposes only under the principle of nominative fair use.

Contents

Introduction: A Planet at a Tipping Point

The Fading Symphony: A Planet Whispering for Help

Listen closely. Can you hear it? It's the sound of a world growing quieter. Imagine walking into a forest that has forgotten how to sing, or gazing upon a reef that has lost its color. This isn't a scene from a dystopian film; it's the reality unfolding around us. The intricate symphony of life, a composition millions of years in the making, is fading.

We are living through an astonishing moment of loss. Since 1970—within a single human lifetime—the populations of wildlife we monitor have plummeted by an average of 73%. Let that number sink in. It's not just a statistic; it's an emptying of our world. It's the silent flight of birds that will never return, the ghostly shadow of creatures that no longer roam. We are standing in the midst of the Earth's sixth great extinction event, but this time, there is no asteroid, no ice age. The architect of this silence is us.

Human activity has redrawn the map of our planet, scarring three-quarters of its land and two-thirds of its oceans. In this vast, altered landscape, an estimated one million species—our fellow travelers on this small blue dot—are now whispering their goodbyes, pushed to the edge of existence. We are, in essence, burning the Library of Life, tearing out pages before we've even had a chance to read them.

The Architects of Absence

This quietening isn't random; it's the direct echo of our choices.

Across the world, we see the same patterns, the same relentless pressures. We clear ancient forests and drain vital wetlands, replacing them with seas of soy and corn. In the heart of Brazil, the Amazon, our planet's lungs, gasps for air under the weight of cattle ranches. In the United States, the great prairies have been fragmented into a patchwork quilt of industrial agriculture.

We hear the ghost of the chainsaw in the rainforests of Indonesia and the Democratic Republic of Congo, where the demand for palm oil, paper, and charcoal leaves behind a hollowed-out landscape. We see the sheen of oil poisoning the lifeblood of Nigeria's Niger Delta and the silent creep of nitrogen from fertilizers choking the rivers of Germany. Our oceans are slowly suffocating in a shroud of plastic.

And over it all hangs the fever of climate change. It is bleaching the life from Australia's Great Barrier Reef, one of the wonders of the natural world, turning it into a ghostly boneyard. It is melting the ancient ice of the Canadian Arctic, pulling the rug out from under an entire ecosystem. Our expanding cities and sprawling infrastructure slice through nature, severing the lifelines that connect wild spaces. Even war and poverty act as accelerants, as seen in Colombia, where the scars of conflict are etched into the land, or in Madagascar, where people are forced to sacrifice unique, irreplaceable forests just to survive.

The Unpayable Debt

For too long, we've treated nature like a free and infinite supermarket. We've taken without thinking of the cost, but the bill is coming due. The truth is, our entire global economy is built on a foundation of natural services we take for granted: the bee that pollinates our crops, the forest that cleans our air, the ocean that feeds our cities.

More than half of the world's GDP—a staggering $44 trillion—is fundamentally tied to the health of our planet. When these natural systems falter, our economic systems will inevitably follow. The World Bank warns that if we allow key ecosystems to collapse, we could see our global economy shrink by $2.7 trillion every single year by 2030.

This isn't about choosing the economy over the environment. That is a false choice. We are learning the hard way that a dying planet cannot support a thriving society. Investing in nature isn't charity; it's the most critical economic insurance policy we could possibly buy.

A Blueprint for Harmony

But in this gathering dusk, a beacon of hope was lit. In late 2022, the world came together to sign the Kunming-Montreal Global Biodiversity Framework—a landmark peace treaty with nature. It's our shared promise to turn the tide, a global roadmap to not just halt the destruction, but to begin healing our planet.

At its heart is a bold and beautiful idea known as Target 3, or "30x30." It is a pledge to protect and heal our world by formally safeguarding at least 30% of our planet's land and sea by 2030. This isn't about drawing lines on a map and walking away. It's about creating living, breathing sanctuaries that are managed with wisdom, respect, and a deep partnership with the Indigenous Peoples and local communities who have been the original guardians of these lands for millennia.

From the Amazonian rainforests of Brazil to the vast wilderness of the United States, nations have embraced this goal. It's a recognition that our survival depends on giving nature the space it needs to recover and thrive once more.

Postcards from a Planet in Flux

A journey across the globe reveals a mosaic of shared struggles and unique challenges:

In Latin America, the Earth's green heart—the Amazon—beats under constant threat. In countries like Brazil, Colombia, and Ecuador, the fight is on to save this global treasure from the relentless expansion of agriculture and mining. The region's staggering 95% decline in wildlife populations is a desperate cry for help.

In North America, the crisis is quieter but no less severe. The vast industrial farms of the U.S. and the creeping impacts of climate change on Canada's boreal forests and tundra show that even the wealthiest nations are chipping away at their natural foundation.

In Africa, the story of humanity and nature is deeply intertwined. From the great apes of the Congo Basin to the fantastical creatures of Madagascar, life is threatened by a complex web of poverty, conflict, and the global demand for resources.

In the Asia-Pacific, a region of explosive growth, the race for progress is often leaving nature in its wake. The orangutans of Indonesia and Malaysia are losing their rainforest homes to palm oil, while the vibrant marine life of the Philippines is threatened by pollution and destructive fishing.

In Europe, even ancient, developed landscapes are showing signs of strain. Spain is battling desertification, while Germany confronts the legacy of industrial pollution, proving that no nation is immune.

Our Choice, Our Legacy

The evidence is overwhelming, and the message is simple: We are part of nature, not apart from it. The health of our planet is our health. The wealth of our planet is our wealth.

The Kunming-Montreal Framework gives us the plan. The 30x30 initiative gives us a tangible goal. But plans and goals are just ink on paper without the will to act. What we need now is a profound shift in our hearts and minds—a move away from a culture of extraction and toward a culture of stewardship. It means changing how we farm, how we consume, how we build our cities, and how we measure success.

The script for the future is not yet written. The final notes of Earth's symphony have not yet been played. We are standing at a tipping point, a moment of choice that will define our legacy. Will we be remembered as the generation that stood by and let the music die? Or will we be the ones who learned, finally, to listen to our planet's whispers and conduct a new, more beautiful symphony of life?

Part I: The Americas - Deforestation, Extraction, and Policy Frontiers

Brazil - Deforestation in the Amazon: The Tipping Point of a Global Ecosystem

A Ghost in the Green

Listen. Can you hear it? Beneath the chainsaw's whine and the crackle of distant fires, there is a new sound rising from the Amazon. It is the sound of a great lung struggling for breath. It is the sound of a living god beginning to forget its own name.

We in Brazil live inside a waking contradiction. We are a people born of the forest, our culture a reflection of its impossible green, our very identity tangled in its roots. Yet, we are the ones holding the blade. The Amazon is not just a collection of trees; it is the planet's memory, a biological epic written over millennia. Its health dictates the rhythm of rain for a continent and steadies the climate for us all. And we are tearing out its pages for profit. We are caught in a fever dream of an economy that pays us to dismantle our own life-support system, trading a cathedral of life for a cattle ranch, a symphony of species for a sterile field of soy. This is more than a paradox; it is a kind of madness.

The Shape of the Wound

To talk about the damage is to speak in a language of ghosts. Forget the statistics; they are just numbers that help us sleep at night. Instead, picture a wound in the Earth so vast it has changed the color of the world map. Imagine a shadow the size of France spreading across the green, a creeping emptiness where life once thrived. This is not a slow decline. It is a hemorrhage.

Scientists say we have erased nearly a fifth of this world. Another fifth is so broken, so scarred and fragmented, it has become a zombie forest—standing, but no longer truly alive. This is not just the felling of timber. It is the silencing of a billion conversations. The Amazon holds, in its living embrace, one in ten of every life form we have ever known. Every acre that falls is a library of unread stories, of potential cures for our diseases, of creatures of impossible beauty, burned before we even gave them a name.

This unraveling is pushing the entire system toward a precipice. A point of no return. The scientists call it the tipping point, a clinical term for a planetary catastrophe. They warn us this edge is near—somewhere between 20% and 25% of total loss. We are standing on the brink, peering into an abyss of our own making.

To cross that line is to trigger a final, irreversible cascade. The Amazon, a system so powerful it creates its own weather, will lose its ability to make rain. The great canopy, which exhales rivers into the sky, will thin and tear. The air will parch. The rainy seasons will stutter and shrink, while the dry seasons will grow into long, fire-choked sieges. The world's greatest rainforest will begin its inexorable transformation into a dry, brittle savanna. It will be a slow-motion death that will unleash a death-rattle of carbon into the atmosphere, convulsing the global climate and executing a mass extinction event of biblical proportions.

A Pink Omen in the Water

If you want to understand this loss, do not look at a chart. Look at the boto, the pink river dolphin. This creature is a myth made flesh, an intelligent, gentle soul that has navigated the Amazon's dark waters since time began. Its health is the river's health. Its fate is our warning.

In the Mamirauá Reserve, a place designed to be a heaven on Earth, the dolphin's world collapsed. Their numbers cratered, a 65% freefall in two short decades. It was a silent scream we failed to hear. Then, in the suffocating heat of late 2023, the scream became visible. The world watched as Lake Tefé, a tributary of the great river, became a tomb. The water, superheated by an unprecedented drought and our warming world, literally boiled the life from them.

Over 330 of their bodies were pulled from the shallows. Their beautiful pink forms, a ghastly and heartbreaking sight, were a message from the dying heart of the ecosystem. They didn't die of old age or disease. They died of heat. They died because their world had been rendered uninhabitable. Their bodies were an omen, a visceral portrait of a system pushed past its breaking point.

The Heart Attack of a Continent

The Amazon is the heart that pumps water to an entire continent. This is not poetry; it is physics. Every day, billions of trees act as colossal biological straws, pulling moisture from the earth and breathing it into the atmosphere in a process called evapotranspiration. This creates an invisible, sky-borne ocean—the "flying rivers."

This great river in the sky, carrying more water than the Amazon River itself, flows west, hits the wall of the Andes, and is forced south. It delivers the rain that sustains the agricultural heartlands of southern Brazil, Paraguay, Uruguay, and Argentina. The food on 300 million plates depends on this celestial pump.

Every tree that falls, every hectare that burns, weakens the pump. It slows the heart. The flying rivers are thinning, faltering. The ever-more-frequent and punishing droughts that plague South America are the symptoms of this weakening pulse. The fate of a farmer in the Argentine Pampas, staring at his cracked and barren soil, is being written by the sound of a chainsaw a thousand miles away in Brazil. To kill the Amazon is to knowingly engineer a future of thirst and hunger for ourselves.

A Desperate Hope on a Knife's Edge

As we stand here in the middle of 2025, the story is one of terrifying suspense. For the first time in what feels like an eternity, a fragile shoot of hope has pushed through the scorched earth. Early this year, data showed that the rate of destruction had plummeted by over 30%, a nine-year low. This was not a miracle. It was the result of will. A new government drew a line in the sand, empowering those who protect the forest and hunting those who burn it.

But this progress is a house of cards. The machinery of destruction—the powerful agribusiness lobby, the political forces that profit from ruin—has not been dismantled. It has only been paused. It waits, patiently, for our will to falter. While our gaze was fixed on the Amazon, the Pantanal wetlands, a paradise of jaguars and giant otters, suffered the worst fires in its history in 2024. The sickness is not in one organ; it is in the body of the nation.

And now, a storm gathers. A bill, a simple plan to pave a road, threatens to be the final, deliberate plunge of the knife. The BR-319 is an old, forgotten scar of a highway, a muddy track that cuts through the untouched heart of the central Amazon. To pave it would be to create a main artery for destruction.

A paved road in the jungle is a catalyst. From it, a "fishbone" pattern of illegal roads will spread like a cancer, opening up millions of acres of pristine forest to a chaotic and unstoppable flood of loggers, miners, and land thieves. Scientists have modeled it. Their conclusion is a death sentence. Paving the BR-319, a single project, could unleash enough destruction to shove the entire Amazon ecosystem over its tipping point. One road could be the act that kills a god.

So this is where we are. In a moment of profound tension, caught in a desperate tug-of-war between a fragile, resurgent hope and the cold, hard logic of the bulldozer. The recent progress proves that the Amazon's eulogy is not yet written. But the shadow of the BR-319 reminds us just how quickly, and how easily, it could be. The fate of the world's living heart, and the stability of our own future, hangs right now between a glimmer of green and the paved road to ruin.

Listen.

If you stand at the edge of the world's greatest forest, you can hear it breathing. It's a quiet, constant hum of life—a billion creatures, a trillion leaves, all inhaling the sun and exhaling the very air we need to survive. But listen closer, and you'll hear another sound. A dissonant, mechanical shriek that tears through the harmony. It's the sound of a chainsaw. And after it comes a more terrifying sound: silence.

The story of the Amazon's demise isn't a natural disaster; it's a human story. It's a tale of a slow, deliberate amputation, driven by a series of choices that mistake a quick profit for lasting prosperity. This is not an abstract environmental issue. It's a crisis of imagination, a failure to see the living, breathing cathedral for the timber.

The Tyranny of the Hamburger

At the heart of the fire and the fallen trees is a simple, voracious appetite. The single greatest force erasing the Amazon from the map is cattle ranching. The numbers are so staggering they feel unreal: for every four ancient trees felled in the Brazilian Amazon, three are sacrificed to create pasture for cows. This isn't a footnote in the story of deforestation; it is the headline.

This hunger has global reach. The demand for beef has made cattle ranching the leading cause of tropical deforestation worldwide. It's a planetary fever. The process is often cloaked in a clever deception known as "cattle laundering." Imagine a shell game played with lives and landscapes: cattle are raised on illegally razed land, fattened in the ashes of irreplaceable forest. Then, just before sale, they are moved to a legitimate, "clean" ranch. Their criminal origin story is washed away, and they enter the global market with a clean bill of health, their connection to the scorched earth conveniently forgotten.

This system thrives on our desire for cheap meat and our distance from the source. Brazil, the world's beef-exporting king, presides over an industry that appears as a pillar of the economy. But its foundation rests on a flawed and brutal calculus—one that sees a living, breathing forest as nothing more than an obstacle between a calf and a slaughterhouse.

The Silent Partner

Beef may be the king of this destructive kingdom, but soy is its powerful duke. While not as direct, its impact is a critical part of the domino effect. The vast, sterile fields of soy blanketing former forest land are not, for the most part, feeding people directly. They are feeding the world's livestock—the chickens and pigs that also end up on our plates. Soy expansion acts like a relentless tide, pushing cattle ranchers, the shock troops of deforestation, ever deeper into the Amazon's heart.

A fragile truce was attempted. The 2006 Amazon Soy Moratorium was a landmark moment, a line drawn in the soil where major corporations agreed to stop buying soy grown on newly deforested land. It proved that a different way was possible, that supply chains could be scrubbed clean. But a truce is not a peace treaty. The global appetite for animal feed continues to swell, and the pressure on this fragile agreement is immense. It remains a constant battle, a testament to how hard it is to tame an economy built on consumption.

A Frontier Ruled by Saw and Shadow

Perhaps the most chilling truth is that this destruction is overwhelmingly against the law. This isn't a tragedy of accidents; it is a crime scene. An estimated 95% of all agricultural deforestation in Brazil is illegal. This isn't a flaw in the system; it is the system.

This rampant illegality is a symptom of a deeper sickness: a vacuum of governance where the law of the land is replaced by the law of the chainsaw. Corruption allows officials to look the other way, while weak enforcement emboldens those who would seize public land, a practice known as grilagem. This is organized crime on an ecological scale, forging the Earth's signature on fraudulent documents to turn a shared inheritance into a private fiefdom. This lawlessness doesn't just kill trees; it fuels violent conflicts, especially against the Indigenous peoples who are the forest's ancestral guardians.

The Devil's Bargain

The argument has always been the same: this destruction is the necessary price of progress. Proponents paint a picture of jobs, revenue, and a nation rising on the back of its natural resources. The mining trucks, the cattle herds, the soy fields—these, they claim, are the engines of economic growth.

It is a beautiful and persuasive mirage.

When you look closer, the illusion shatters. This model of "development" is a devil's bargain, offering fleeting gains for a handful of powerful players while handing the devastating long-term bill to everyone else. It's an economic model that can count the price of everything but the value of anything.

A stunning 2023 World Bank report finally did the math that matters. It didn't count the timber and the beef; it calculated the value of the living forest. It tallied the cost of the "ecosystem services"—the rain the Amazon generates for farms across the continent, the carbon it stores, keeping our climate stable, the medicines held in its genetic library.

The verdict was an economic bombshell: continuing down this path of destruction will cost the Brazilian economy $317 billion in losses every year.

This isn't an environmentalist's plea; it's a banker's warning. The profit from razing an acre of forest is a pittance. The report found that the losses from destroying those free services are seven times higher than the profits gained from the commodities that replaced them. Cattle ranching yields a meager $75 per hectare. Sustainable activities like agroforestry or harnessing the forest's native products could generate $108 per hectare. We are not just liquidating our most valuable asset; we are doing it for a fraction of its worth. It's like burning down a priceless library to sell the ashes for pennies.

The Echoes in a Silent Forest

The final lines of this ledger are written in human suffering. When a tree falls in the Amazon, it does not fall in silence. It echoes in the lives of the Indigenous communities, the rubber tappers, and the river dwellers whose cultures are woven from the forest's fabric. Their identities, their food, their pharmacies, and their spirits are intrinsically tied to the land. For them, deforestation is not an economic policy; it is an existential threat.

The violence that erupts on this frontier is a battle for survival. Environmental defenders and Indigenous leaders are murdered for protecting lands that are legally theirs. And what of the grand promise of prosperity for the region? After 50 years of relentless clearing, which has erased a fifth of the original forest, the Amazon region's contribution to Brazil's GDP has remained stagnant at a paltry 8%.

This is the ultimate indictment. The destruction has not brought wealth to the many. It has concentrated it in the hands of a few, leaving a legacy of social conflict, deepening inequality, and a land stripped of its ability to provide for its people.

The Choice

The story of the Amazon is the story of a choice. We have been trapped in a fever dream of extraction, believing that we could build a future by dismantling our life support system. But the numbers are in, the human cost is clear, and the planet is sending us the bill.

The alternative is not to turn the Amazon into a museum. It is to awaken to a new dream of stewardship, one that recognizes the staggering genius of a living forest and builds an economy from that genius—an economy of sustainable harvests, of ecotourism, of scientific discovery, of life. This path offers not just a healthier planet, but greater and more equitable prosperity.

We are standing at a crossroads. One path leads to a barren inheritance. The other, to a living legacy. The Amazon is asking which future we will write.

In the heart of the world, there is a nation at war with itself. Brazil, a country whose name is synonymous with the emerald expanse of the Amazon and the vibrant cacophony of life it holds, is caught in a profound struggle for its own soul. This isn't a battle fought with armies, but one waged in courtrooms, on the world stage, and most critically, in the stark, unforgiving numbers of its own bank ledgers. To understand Brazil today is to understand a symphony of contradiction.

The Gavel and the Seed

Imagine a silence descending upon a government, a deliberate quiet in the face of roaring chainsaws and encroaching smoke. This was the vacuum Brazil's Supreme Court shattered. In a move that was less a legal judgment and more a moral awakening, the Court confronted this political paralysis head-on. Through landmark rulings, it gave a name to the government's inaction: an "unconstitutional omission."

Think of what that means. The Court declared that the failure to protect the environment was not a passive oversight, but an active violation of the people's constitutional right to breathe clean air, to drink clean water, to exist in a balanced world. It was a revolutionary idea: that a government could break the law by doing nothing at all.

This wasn't just theory; it had teeth. The Court ordered the state to open its clenched fists and release the lifeblood for conservation: the Amazon Fund and the National Climate Fund. These vast reservoirs of money, bankrolled by international partners and domestic commitment, were designed to be the financial muscle behind protecting the rainforest. They had been frozen, gathering dust while the forests burned. The Court's order was like a jolt of adrenaline to a slumbering giant, forcing the reactivation of the very mechanisms meant to ensure its survival. It was a powerful demonstration of the judiciary acting as the nation's conscience, a guardian ensuring that the promise of a healthy future wasn't just ink on paper, but an enforceable, living right.

A Promise Renewed on the World Stage

Outwardly, Brazil began to dance a new dance. After a period of turning its back on the global community, the nation stepped back into the spotlight, its voice ringing with renewed commitment. It was a slow turn, but in early 2025, Brazil formally embraced the Kunming-Montreal Global Biodiversity Framework, the world's most ambitious pact to save nature.

At the heart of this pledge is the monumental 30×30 target—a global vow to protect at least 30% of the Earth's land and sea by 2030. For a megadiverse nation like Brazil, this is a promise of colossal proportions, a commitment to safeguard vast stretches of rainforest, sprawling wetlands, and thousands of miles of vibrant coastline.

This isn't just talk. The promise is attracting powerful allies, like Bloomberg Philanthropies, which has channeled funds specifically to help shield Brazil's precious marine ecosystems. The timing is electric. With the world's gaze turning towards Belém, an Amazonian city, for the pivotal COP30 climate summit in 2025, Brazil has handed itself the microphone. It is a golden opportunity to transform from a subject of international concern into a beacon of environmental leadership, with the living, breathing Amazon as its witness. But a vow made on the world stage is one thing; keeping it in the muddy fields back home is another entirely.

The Poisoned Well of Finance

Here, in the cold, hard numbers of the national economy, the beautiful story begins to unravel. This is the central paradox of Brazil's environmental struggle. While one arm of the state reaches for global accolades, the other is quietly feeding the beast it claims to be fighting.

The problem is twofold. First, there is a massive biodiversity financing gap—the country simply doesn't have enough money allocated to do the job right. But the more sinister issue is where the existing money flows. A stunning World Bank analysis revealed that nearly half—a staggering 46%—of all corporate loans from Brazilian banks are given to industries whose survival is intensely dependent on the very ecosystems being destroyed. Think of agriculture, forestry, and fisheries all drinking from a well they are simultaneously poisoning. As nature degrades, these businesses will falter, their loans will sour, and the entire financial system will tremble. The health of the rainforest and the health of the economy are not separate; they are one and the same.

And here is the most damning truth: the government itself is holding the hose. The Brazilian Agricultural Plan, or Plano Safra, is a massive program that injects billions of dollars in subsidized credit to the nation's farmers. It is intended to grow food and support rural communities. Yet an investigation found that between 2020 and 2022, nearly a third of all rural properties that illegally cleared forests had been rewarded with this cheap, state-sponsored money.

It is a policy at war with itself, a nation paying an arsonist with one hand while funding the fire department with the other. This mirrors a global tragedy of nearly $542 billion in harmful agricultural subsidies that actively degrade the planet.

So Brazil stands at a crossroads, a nation of breathtaking contradictions. It holds a judge's gavel that can command the protection of a forest, and it signs international vows to heal the Earth. But it also holds a checkbook that finances destruction. The path forward demands more than just good intentions; it demands coherence. It requires Brazil to stop funding its own demise and to align its immense financial power with its ecological soul. The world watches, not just to see if the Amazon can be saved, but to learn which Brazil will ultimately write the final chapter of its story.

Colombia - Peace, Land Use, and Illicit Economies: Navigating a Complex Conservation Landscape

In the heart of Colombia, a profound silence is being broken. It's not the echo of the 50-year war that haunted its jungles and mountains, but the roar of chainsaws and the groan of earthmovers. Here, in a land that holds ten percent of the world's living soul within its borders, peace has come with a devastating paradox: the quiet of the guns has, in many places, become a death sentence for the trees.

Imagine a country stitched together from ecosystems so vibrant they feel mythical. The Amazon, an emerald ocean of life. The Andes, whose misty cloud forests cling to the slopes like secrets. And rising above them, the Páramos—the nation's water-bearing soul, high-altitude moorlands that act as celestial sponges, giving drink to more than three-quarters of the population. This is the Colombia that breathes life into the world.

But this living masterpiece is coming undone. The country's tropical dry forests are nearly ghosts, with 95% gone. The Andean forests have been stripped back by 70%. For species found nowhere else on earth, like the electric-blue-beaked Curassow or the whimsical Cotton-top Tamarin, this isn't a statistic; it's an eviction notice from the only home they have ever known. Their existence is unraveling thread by thread.

The Amazon bears the deepest scars. Here, a great "arc of deforestation" smiles grimly across the landscape. The numbers are dizzying, swinging wildly like a national mood. A 35% surge in forest loss in 2024, driven by the shadow economies of armed groups. A supposed 33% drop in early 2025. Who can you believe? While the government claims victory, independent eyes see a hemorrhage, with nearly 89,000 hectares of protected forest vanishing in just six months. The truth is, in a country where the state's shadow is faint in the wildest places, the full story is a puzzle, its pieces scattered across the jungle floor.

The Ghosts of War and the Hunger for Land

To understand why, you must understand the strange peace. For decades, the FARC rebels, hidden beneath the dense canopy, acted as an unintentional green wall. Their need for cover from the sky inadvertently protected the forests from the march of progress and plunder. But when they laid down their arms in 2016, they left behind a vacuum of power—a void that nature abhors and men with ambition rush to fill.

Into this void stepped dissident factions, criminal syndicates, and land speculators. The forest became the prize. Clearing it was the quickest way to stake a claim, to turn a living ecosystem into a dead asset on a map.

The cow became the new soldier in this war for land. Cattle ranching, a sprawling, relentless force, is the single greatest driver of this destruction. It's a simple, brutal calculus: a cleared pasture is seen as "productive" land, a way to launder money, or simply the only path to a livelihood for a family with no other options.

Elsewhere, a different kind of hunger gnaws at the earth: the fever for gold. Illegal mining carves up the land and poisons the rivers with mercury, a silver toxin slithering through the nation's veins. It chokes waterways, kills fish, and leaves a legacy of contamination that will haunt generations. And woven through it all is the ghost that will not rest: the coca leaf. While its direct footprint may be smaller, the drug trade carves clandestine roads and airstrips, pushing ever deeper into untouched frontiers, a "balloon effect" where squeezing one area only causes another to swell.

A New Peace Treaty: This Time, With Nature

Yet, to cast Colombia solely as a victim is to miss the most thrilling part of its story. This is a nation refusing to surrender its future. Facing a crisis born of its own complex history, Colombia is becoming a global architect of hope, forging a new kind of peace treaty—this time, with the Earth itself.

It is here, amidst the turmoil, that a radical idea is taking root: teaching money to speak the language of the forest. Colombia is pioneering financial instruments that sound more like science fiction than banking. They've issued biodiversity bonds, attracting global capital not for concrete and steel, but for the regeneration of forests and the healing of mangroves. They are "greening" the very loans that fuel their agricultural sector, creating financial pathways that reward farmers for raising cattle or growing coffee in harmony with the land, not at its expense.

This is a nation turning its wounds into wisdom. Its decision to host the world's most important biodiversity summit, COP16, under the banner of "Peace with Nature," is a powerful declaration. It's Colombia stepping onto the global stage, not to ask for help, but to lead. It is offering its own painful, paradoxical experience as a lesson: you cannot build lasting human peace on a foundation of ecological war.

The country is championing a global deal for fairness, demanding that the world's wealthy nations, which benefit from the genetic codes hidden in Colombia's biodiversity, share the profits equitably. And it is leading a coalition to secure real, hard funding to protect the planet's life-support systems.

Colombia's story is a microcosm of our planet's own struggle. It is a breathtaking, heartbreaking, and fiercely hopeful testament to the fact that the path to a thriving human future is inextricably bound to the well-being of the natural world. In the dance between destruction and innovation, Colombia is trying to teach itself, and the world, how to finally make peace with nature, so that the silence that follows is not one of loss, but one of life restored.

Colombia's green heart is bleeding.

It is a slow, deep hemorrhage, not of blood, but of life itself. In this land of impossible biodiversity, where orchids burst from ancient trees and the air hums with the wings of a thousand different birds, a profound sadness has taken root. The assault on its natural soul is not a simple crime, but a complex sickness, born from the ghosts of war and the ravenous appetite of a shadow economy. To understand the crisis is to understand a story of tragic ironies and human desperation.

The Scar of Peace

For half a century, a brutal, unintended quarantine sealed off vast expanses of Colombia. The FARC guerrillas, while a plague upon the people, were also the reluctant guardians of the wilderness they occupied. Their rule of terror was a fortress wall, keeping the machinery of industrial extraction at bay. In their shadow, the jaguar still prowled and ancient forests breathed, oblivious to the world outside.

Then, in 2016, a whisper of hope became a national sigh of relief. Peace. The war was over. But as the FARC laid down their arms and melted back into society, the silence they left behind was not one of tranquility. It was the silence of a void. This vacuum became a vortex, sucking in a new legion of destroyers. Splinter militias, land barons, and criminal prospectors rushed into the unguarded Eden, and the peace accord, meant to heal the nation, inadvertently tore open a new wound—this time, on the land itself.

The view from space became a chronicle of this tragedy. Green canopies that had stood for generations vanished in months, replaced by the raw, brown scar of exposed earth. The peace, so desperately won, had unleashed a new kind of war, a silent, relentless war against the wild.

The Agents of Decay: A Fever in the Land

The forces consuming Colombia's natural heritage are not separate illnesses, but symptoms of a single, raging fever.

The Cow as Conqueror

The single greatest destroyer is not a machine, but an animal: the cow. Cattle ranching is the vanguard of deforestation, a relentless wave that turns forest into pasture. But this is about far more than beef. In the fractured legal landscape of rural Colombia, a cleared field is a statement of ownership. To fell the forest and fence the land is to forge a claim, to turn a public commons or an ancestral territory into private property.

This is the insidious art of the land grab, often orchestrated by powerful criminal syndicates. They pour illicit fortunes into vast cattle operations, not for the ranching itself, but to launder money and speculate on the land. The forest is razed, the land's value skyrockets, and the profits are funneled back into the shadows. The cow becomes a conqueror, its lowing the only hymn sung over stolen ground, while Indigenous and Afro-Colombian communities are driven from the lands that have cradled their cultures for centuries.

The Poison in the Veins

Deep in the newly opened territories, illicit economies act as a shadow government, writing laws with bullets and bribes.

The White Blight of Coca: The dream of replacing coca with legal crops has been a difficult one. As power shifted, so did the coca fields, creeping deeper into national parks and protected reserves. Each new field is a patch of stolen forest, and the chemicals used to process the leaf bleed into the soil and streams, a poison seeping through the land's veins. It is a tragic dance: communities, trapped between coercion and poverty, plant the leaf, only to have it eradicated, pushing them to clear new ground elsewhere in a desperate cycle of survival.

The Golden Curse: A different kind of fever glitters in the rivers: the fever of illegal gold mining. This is a brutal, chaotic industry. Miners tear into riverbeds, ripping apart the aquatic world in a frantic search for treasure. They use mercury, a liquid silver death, to bind the gold. This toxin is then flushed into the water, a permanent poison that works its way up the food chain—from fish to the humans who eat them. In the communities downstream, the mercury becomes a ghost in the blood, causing neurological damage and birth defects. It is an economy that buys a few grams of gold at the cost of a river's life and a people's future.

The Illicit Arteries

None of this would be possible without the roads. Not highways, but clandestine dirt tracks, carved into the heart of the jungle. These illegal roads are the arteries of destruction. Once cut, they carry the infection deep into the pristine wilderness, opening the door for loggers, miners, ranchers, and growers.

From these arteries flows the lifeblood of the illegal timber trade. Nearly half of all wood sold in Colombia is stolen from its forests. Prized trees are felled, their absence leaving a wound in the canopy, and shipped across the world to markets in China and India. The forest's bones are sold to the highest bidder, funding the very corruption that allows the destruction to continue.

The Human Cost of a Dying Land

The collapse of these ecosystems is not an abstract environmental problem. It is a human catastrophe, and its weight falls heaviest on the shoulders of the most vulnerable.

The Mirage of Progress

A dangerous myth persists: that clearing the forest brings prosperity. The reality is a cruel inversion. As the trees fall, so do the fortunes of the people. The short-term profits are siphoned away by outside powers, leaving local communities with degraded land, polluted water, and lost resources. The forest, with its gifts of clean water, stable soil, and potential for sustainable economies like eco-tourism, is their true natural capital. By destroying it, these regions are not developed; they are trapped in a deeper, more permanent state of poverty.

A Nation's Thirst

The great watersheds, the sources of Colombia's water and energy, are being choked. In basins like the Upper Sinú, deforestation has stripped the land of its ability to hold water. The soil bleeds into the rivers, silting up reservoirs and crippling the hydroelectric dams that power the nation. The land, losing its grip, can no longer buffer against nature's extremes, unleashing ruinous floods in the wet season and punishing droughts in the dry. The loss of these natural defenses is a multi-million dollar blow to the economy, a direct threat to the water flowing from the taps in cities and the energy that lights the country.

A War for Water and Earth

To defend nature in Colombia is to risk your life. The country is a global hotspot for environmental conflict, a place where standing up for a river or a forest can get you killed. Community leaders, Indigenous guardians, and local activists who dare to oppose the land grabbers and illegal miners are met with threats and violence. They are the front-line soldiers in this war, their silenced heartbeats a grim testament to the stakes. Their struggle is not just for land, but for the soul of their communities and the future of a livable planet.

Fleeing the Fever

Finally, the degraded land is beginning to expel its own people. Climate change, amplified by deforestation, is turning weather events into catastrophes. Landslides sweep away villages; floods drown farms and livelihoods. This intensified chaos is now a major driver of forced migration. Families are uprooted, becoming environmental refugees in their own country, fleeing a land that can no longer sustain them.

The story of Colombia is the story of a nation at a crossroads. Its immense natural wealth is also its greatest vulnerability. Healing this bleeding land requires more than just environmental policy; it requires a profound reckoning with the forces of violence, inequality, and corruption. It demands a new dream for the nation, one where peace means not an open door for destruction, but a sacred pact to protect the vibrant, living heart of Colombia for all time.

In the heart of Colombia, where the emerald canopy of the Amazon meets the sky, a quiet rebellion is underway. This is not a rebellion of arms, but of ideas. Faced with the relentless bleeding of deforestation, a wound that scars its most vital ecosystem, the nation is refusing to be a victim of fate. Instead, it has become a living laboratory, a place where the future of conservation is being forged in a crucible of law, diplomacy, and radical financial imagination. This is the story of how a country is weaving a lifeline for its natural heritage, thread by powerful thread.

I. The Conscience of the Courts and the Fist of the State

Colombia's defense of its environment begins not in the forest, but in the hallowed halls of justice. It's a strategy that gives the law a voice, and nature a seat at the table.

A. When the Amazon Itself Went to Court

In 2018, a legal earthquake shook the nation. In an act of breathtaking courage, a group of 25 young Colombians—children and young adults—sued their own government. Their claim was profound in its simplicity: by failing to stop the destruction of the Amazon, the state was violating their most fundamental rights to life, to water, to a future.

In a verdict that echoed around the world, the Supreme Court of Justice agreed. It did something unprecedented: it declared the Colombian Amazon a legal entity, a subject with its own rights to be protected, respected, and restored.

This was more than a ruling; it was a philosophical shift. The forest was no longer just a resource to be managed, but a living being with legal standing. This decision, born from the conviction of its youngest citizens, handed conservationists a powerful legal sword. It transformed environmental protection from a political preference into a constitutional duty. While the journey from this landmark decree to stopping every chainsaw on the ground is long and fraught with challenges, the message is clear: in Colombia, the courts can be the conscience of the land.

B. The Camouflage Conundrum

Contrasting with this legal elegance is the state's more blunt instrument: a militarized war on environmental crime. Here, the strategy is one of force. Soldiers and police deploy into the jungle, dismantling illegal mines, destroying coca labs, and intercepting illicit timber—all major arteries of deforestation. The logic is clear: these are not just environmental issues; they are matters of national security, often entangled with armed groups and criminal syndicates.

Yet, this approach walks a razor's edge. Critics argue with compelling force that it often punishes the symptoms while ignoring the disease. The farmer clearing a small patch of land out of desperation—a pawn in a much larger game—is an easy target. The powerful, shadowy figures who orchestrate the devastation from afar remain untouched. This "boots on the ground" strategy can feel like a game of whack-a-mole, displacing illegal activity from one region to another without addressing the deep-seated poverty and lack of opportunity that fuel it. It raises a difficult question: can you truly save a forest by criminalizing the people who live in its shadow? Many believe the only true victory lies in a strategy that pairs handcuffs for criminals with a helping hand for communities.

II. An Ambassador for Biodiversity

Colombia has carried its fight beyond its borders, taking the microphone on the world stage to champion the cause of all life on Earth.

A. The Cali Consensus

This global leadership was on full display when Colombia hosted the UN Biodiversity Conference (COP16) in Cali in late 2024. By welcoming the world, Colombia positioned itself not as a supplicant, but as a leader shaping the global conversation. The summit's outcomes were a testament to this influence. A key achievement was the creation of the 'Cali Fund,' a groundbreaking mechanism to ensure that the wealth generated from an organism's genetic data—its Digital Sequence Information (DSI)—is shared fairly with its country of origin. It's a move toward digital justice for the planet's most biodiverse nations.

Furthermore, the conference gave new power to the planet's oldest guardians. A new work program was established to deeply embed the wisdom and rights of Indigenous Peoples and Local Communities (IPLCs) into the heart of global conservation policy. It was a formal acknowledgment that the keepers of traditional knowledge are not stakeholders to be consulted, but partners to be empowered.

B. Turning a Number into a Promise

Colombia is translating global ambition into local action, particularly with the "30x30" target—the global pledge to protect 30% of the planet's land and sea by 2030. This isn't just a talking point; it's a mission. A massive $245 million agreement, backed by a coalition of international partners, has been secured to create a permanent shield around a portfolio of the country's most vital natural areas. It's a financial fortress for biodiversity.

In a remarkable act of solidarity, Colombia is also joining forces with its neighbors, Ecuador and Peru. Using state-of-the-art spatial data, the three nations are creating a shared map of hope, identifying the most critical conservation corridors that span their borders. It's a powerful recognition that ecosystems don't recognize human boundaries, and neither should their protection.

III. The Alchemy of Finance: Turning Capital into Canopy

Perhaps Colombia's most revolutionary contribution is its role as an alchemist, transforming the cold logic of finance into a force for living nature. Recognizing that government coffers alone can never fund the sheer scale of conservation needed, the country is pioneering new ways to make money serve the wild.

A. A Bond with a Green Soul

In a global first, BBVA Colombia, with the backing of the World Bank's International Finance Corporation (IFC), issued a biodiversity bond. This wasn't just another green investment; it was a financial instrument with a green soul, meticulously designed to channel private capital into projects that actively heal the wounds of biodiversity loss—from sustainable farming to regenerative tourism. It was a signal to the world that protecting nature could be a powerful asset class.

This is just one tool in a growing workshop. The UN's BIOFIN initiative is working hand-in-glove with the Colombian government to craft a whole suite of financial solutions. They are building markets for biodiversity offsets, where development's impact in one area funds restoration in another. They are nurturing community-managed conservation funds, putting resources directly into the hands of local stewards. Most critically, they are working to "green" agricultural finance, rewriting the rules of credit so that farmers are rewarded, not penalized, for practices that nourish the soil and protect the forest. It is a quiet revolution, changing the very DNA of how money flows across the landscape.

B. The Surprising Power of a Paycheck

But do these financial incentives work? A fascinating case study on Payment for Ecosystem Services (PES) from the Colombian Amazon piedmont offers a stunning answer. The common fear is that paying people to conserve nature—for example, to protect a watershed—might "crowd out" their intrinsic motivation. The worry is that they will protect the land only for the money, and stop the moment the payments do.

The study in Colombia found the exact opposite. When the PES program ended, the pro-environmental behavior of the community actually increased. The financial payments had acted not as a replacement for conscience, but as a catalyst for it. The funds provided a bridge, allowing communities to experiment with sustainable practices and see the long-term benefits for themselves. The money was the spark, but the community's own growing commitment became the enduring flame. It suggests that financial tools, when designed with wisdom and respect, can do more than buy conservation—they can build a lasting culture of stewardship.

A Future Woven from Hope

Colombia's journey is far from over. The deep roots of deforestation—driven by complex webs of crime, poverty, and global demand—are notoriously difficult to sever. Yet, the nation's story is one of defiant creativity. By braiding together the moral force of its courts, the strategic clout of its diplomacy, and the revolutionary potential of its financial innovations, Colombia is offering more than just a case study. It is offering a blueprint for hope. It is a testament to the fact that confronted with the greatest of challenges, the most powerful response is a fierce, creative, and unwavering love for the land.

Ecuador - Balancing Extraction and Conservation: Oil in Yasuní and the Galapagos Debt-for-Nature Swap

Ecuador is a nation with two souls. One soul is that of a poet, a philosopher. It's the soul that looked upon the Earth—the rivers, the mountains, the ancient forests—and saw not a resource, but a relative. In a move that sent ripples of awe across the globe, this soul wrote its reverence into law, granting constitutional rights to Nature herself. It made a sacred promise, positioning this small country as the world's conscience.

But Ecuador's other soul is that of a survivor, one haunted by necessity. This soul knows the gnawing pang of economic hunger and feels the immense pressure to feed its people. It looks at the same land and sees the black gold of oil and the glittering promise of minerals buried deep beneath the life it holds sacred.

This is the wrenching paradox of Ecuador. It is a nation caught in a painful tug-of-war between its highest ideals and its harshest realities. It is a living, breathing story of our planet's most urgent struggle: the fight to find a way for humanity to thrive without devouring the very home that sustains us.

A Journey Through a Wounded Paradise

To understand Ecuador, you must walk its sacred grounds, beginning where a single idea changed our understanding of life itself.

Step onto the volcanic shores of the Galápagos Islands. This is no mere archipelago; it is an evolutionary cradle, a living museum where Charles Darwin first glimpsed the grand, sprawling story of life. Here, isolation brewed a magical menagerie of creatures found nowhere else on the planet. But this cradle of creation is now under siege. Uninvited guests, brought on human tides, threaten to unravel it all. Goats, ghosts of a pastoral past, march across the islands, chewing the landscape down to barren rock, starving the deliberate, ancient giant tortoises. Fire ants, tiny terrors, swarm in silent, deadly waves, preying on the next generation of birds and reptiles.

Look out at the water, and you'll see the second wave of the assault. The ocean, once a pristine moat, now delivers a confetti of poison. Plastic, our civilization's careless signature, chokes the shorelines. Sea lions, playful and curious, become entangled in the ghostly nets of forgotten fishing fleets. Majestic sea turtles mistake floating bags for food, their bodies discovered later, starved from the inside out. The hammerhead sharks, those magnificent, otherworldly guardians of the reef, have seen their numbers slashed in half since the new millennium, victims of a relentless hunger for their fins. And as our world warms, the very ocean turns against its own. Rising temperatures cook the life from the coral reefs, bleaching them into bone-white graveyards and silencing the vibrant cities of fish they once sheltered.

Now, journey inland, from the sea to the soul of the continent, to a place that represents the pinnacle of earthly life: Yasuní National Park. This isn't just a forest; it is the Amazon's beating heart, a cathedral of biodiversity so rich it defies imagination. Stand in a single hectare—a patch of land smaller than two football fields—and look up. You are surrounded by more tree species than exist in all of the United States and Canada combined. This is a sanctuary for the world's shyest wonders: the jaguar moving like a shadow, the impossible pink river dolphin, and the giant anteater.

But this green Eden holds a deeper secret. It is home to the Tagaeri and Taromenane, Indigenous peoples who live in voluntary isolation, the last free people of the forest. They are the human embodiment of this wild place, their existence a testament to a life woven into the fabric of nature. Their home is legally protected, an "intangible zone." Yet, beneath this sacred ground lies a treasure that could be its doom: crude oil. The battle for Yasuní has become the defining struggle for Ecuador's soul. The nation once begged the world to help it keep the oil in the ground, a radical plea to choose life over profit. The plea went unanswered. Now, despite a historic vote by its people to halt the drilling, the shadow of the oil rig still looms, threatening to shatter the silence of the forest and the lives of its most vulnerable guardians.

Climb higher now, into the Andes, where the mountains breathe clouds. Here lie the Chocó cloud forests, enchanted realms shrouded in mist. Water is the architect of this world, dripping from every surface, nurturing a riot of life. Trees are cloaked in emerald mosses and wear crowns of orchids. This is a land of endemism, a place where evolution, sealed in pockets of mountain fog, has crafted creatures of breathtaking rarity, from the fiery Andean cock-of-the-rock to jewel-toned frogs that exist in a single valley.

This ethereal world is fragile. The blade of the chainsaw and the relentless march of agriculture slice through these misty forests, leaving behind wounds of erosion and silence where there was once a chorus of life.

Where Hope Takes Root

Yet, even in the most wounded places, hope is a tenacious seed. In 2025, amidst the peril, miracles are quietly unfolding.

In a region called Centinela, botanists recently walked through fragments of forest long given up for dead. And there, they found ghosts. A vibrant orange flower, Gasteranthus extinctus, a plant whose very name means "extinct," was found blooming defiantly. Its rediscovery is more than a scientific footnote; it's a whisper from a lost world, a poignant reminder that even the smallest, most forgotten patches of forest can harbor impossible treasures, pleading for a second chance.

In the Chocó, another kind of hope is taking root. Here, local communities are refusing to choose between preservation and survival. They are weaving them together. Through community-led agroforestry, they are learning to grow coffee and cacao in the shade of the native canopy, becoming gardeners of the forest rather than its destroyers. They are proving that a living forest can sustain a community, that people can be its most passionate protectors.

Ecuador's story is a mirror. It reflects the war within all of us—the chasm between what we know is right and what we feel is necessary. The fate of the Galápagos tortoise, the uncontacted tribes of Yasuní, and the ghost flower of Centinela is not just Ecuador's story. It is ours. The coming years will reveal which of its souls—and which of ours—will ultimately win.

In Ecuador, the earth has a fever.

You can feel it in the deep, humid pulse of the Yasuní, a forest so alive it's considered one of the planet's lungs. Here, under a canopy teeming with more life in a single hectare than in all of North America, lies a treasure. But it is not the iridescent flash of a hummingbird's wing or the silent jaguar that men in suits and hard hats seek. They are drilling for a different kind of lifeblood—the black, viscous crude oil that rests beneath the forest floor.

This is the great paradox of Ecuador: a nation haunted by a ghost. The ghost is its colossal foreign debt, a specter that whispers in the ear of every president, demanding payment. And the only offering it seems to accept is the heart of the Amazon itself. To appease creditors in far-off capitals, particularly the immense financial power of China, the country must drill. The logic is as simple as a death sentence: to live today, we must kill a part of our tomorrow.

The process is not a quiet surgery; it is a brutal assault. Imagine a blade slicing through a masterpiece. That is what an access road does to this ancient forest, opening arteries for colonists and illegal loggers to flood in. Picture a vibrant, breathing ecosystem scraped clean, replaced by the sterile steel and mud of a drilling platform. See the pipelines, like black veins of poison, snaking through a world that has only ever known clean water. Each step is a wound, a tear in a fabric woven over millennia, leaving behind scars of deforestation, polluted streams, and a fractured, dying land.

This fever is not felt equally. For the Waorani people, who have walked these forest paths for centuries, it is an invasion. Their home, once a sanctuary governed by the cycles of the sun and the river, has become a warzone.

The oil brings conflict, pitting neighbor against neighbor, community against the state, and a people against the deafening roar of machinery that promises progress but delivers despair.

Deeper in the forest's embrace live the Tagaeri and Taromenane, the last echoes of a world untouched. They are shadows, living in voluntary isolation, a courageous choice to preserve their existence. But the fever is closing in. The roar of the outside world shatters their sacred silence, bringing with it diseases their bodies cannot fight and a violence they cannot comprehend. Their continued existence hangs by a thread, a thread being cut by decisions made in boardrooms thousands of miles away.

There is a cruel joke at the heart of this tragedy. A recent study, a kind of financial autopsy on the Yasuní project, revealed the devastating truth. It followed the money, and found that for every dollar of wealth pulled from beneath the forest floor, a staggering 79 cents flows out of Ecuador. It vanishes into the accounts of multinational corporations and foreign governments. The people of the Amazon are left with less than a quarter of the value, but 100% of the cost. They are left with the poisoned rivers, the empty forests, the broken communities. They are left to live in the wreckage of a party they were never invited to.

And the fever is spreading. Beyond the oil fields of Yasuní, a relentless hunger consumes the rest of the nation's forests. Industrial agriculture fells ancient trees for seas of oil palm, an ingredient hidden in the snacks and soaps of the developed world. In a strange, green twist, the global rush for wind energy has created a voracious appetite for balsa wood, a key component in turbine blades. As the world builds its clean future, it carves it out of Ecuador's forests.

Miners, both legal and not, chase veins of gold, blasting away hillsides and bleeding mercury and cyanide into the rivers. The water that once gave life now carries sickness, and the soil that once fed families turns barren. And everywhere, there are the roads, always the roads, pushing ever deeper, bringing with them a wave of destruction that never recedes.

This is a nation building its future on a foundation of ghosts. Climate change, fed by this very deforestation, promises to wither the nation's farms. By 2050, rice harvests could plummet by 60%, leaving millions to face hunger. The oil economy itself is a ticking clock. As the world weans itself from fossil fuels, the black blood that Ecuador sells will lose its value, threatening to crash an economy chained to a dying industry.

In the end, the story is written on the bodies of the people. It is in the respiratory illnesses of children breathing polluted air, in the desperation of families whose farms and fisheries have been poisoned into oblivion. It is the story of a nation caught in a global economic machine that demands a terrible sacrifice: the sacrifice of its own irreplaceable soul for a prosperity that never quite arrives.

To truly understand Ecuador, you must see it not as a single country, but as a nation with two hearts beating in its chest. One heart is ancient, wild, and green, pulsing with the rhythms of the Amazon and the wisdom of "Pachamama," Mother Earth. The other is modern, relentless, and black, pumping the crude oil that has for decades been the lifeblood of its economy. The story of Ecuador is the story of this profound internal struggle—a nation at war with itself, and a nation inventing its own peace.

In 2008, the green heart roared to life. In a move that sent shockwaves through the world's halls of justice, Ecuador rejected centuries of law that treated nature as mere property. It listened to the whispers of its indigenous peoples and enshrined the Rights of Nature in its constitution. Suddenly, a river was no longer just a resource to be dammed or polluted; it was a living entity with the right to flow. A forest was not just timber waiting to be cut; it was a community of beings with the right to exist. This wasn't poetry; it was law. For the first time, a court could hear a case brought on behalf of a poisoned stream or a clear-cut jungle, giving a legal voice to the voiceless. It was a flare sent up from the middle of the world, signaling that a different relationship with our planet was possible.

But the black heart of the economy beat on, powerful and demanding. This heart had built the roads, funded the schools, and paid the salaries. Its hunger for oil drove drills deep into the earth, often beneath the very sacred grounds the new constitution was meant to protect. This clash of hearts reached its crescendo in the Yasuní National Park, a place so bursting with life it's considered one of the planet's biological crown jewels and home to tribes living in voluntary isolation. Here, beneath a canopy teeming with impossible biodiversity, lay the ITT oil block.

For years, the battle for Yasuní was a quiet war fought by activists and indigenous guardians. Then, in August 2023, the question was put to all of Ecuador in a national referendum. It was a national reckoning. And the people spoke with a single, thunderous voice. Nearly 59% of the nation voted to turn off the pumps. They chose the jaguar over the oil derrick, the uncontacted tribes over the quarterly revenue report. It was a breathtaking act of democratic will, a populace demanding that their government honor the nation's green soul.

Yet, victory's celebration was short-lived. The government, caught between the people's mandate and the terrifying reality of a billion-dollar hole in the budget, has hesitated. Citing overwhelming costs and economic shock, the dismantling of the Yasuní oil infrastructure has stalled. The delay has drawn sharp rebukes from the international community and UN experts, who see a government defying the sacred will of its own people. It's a painful, public lesson in just how deep the addiction to fossil fuels runs, and how agonizing the withdrawal can be, even when the cure has been chosen by popular vote.

But just as this internal conflict seems to define Ecuador, the nation pivots, revealing a stunning capacity for creative genius. In a brilliant display of financial alchemy, Ecuador has shown the world how to make the pragmatic heart serve the green one. In 2023, it executed the largest debt-for-nature conversion in history.

Imagine taking a mountain of national debt—over $1.6 billion—and transforming it into a shield for one of nature's greatest treasures. By cleverly buying back its own debt at a steep discount with the help of international partners, Ecuador freed up a river of money. But instead of flowing into government coffers, that money is now contractually bound to flow into a massive, permanent conservation fund. This single deal will pour an estimated $450 million directly into protecting the enchanted isles of the Galápagos and the vital Hermandad Marine Reserve, creating a protected superhighway for migrating sharks, whales, and sea turtles. It's a masterstroke, turning the cold, hard logic of finance into a warm-blooded promise to the planet.

This isn't a one-off trick. Ecuador is stepping onto the global stage as a teacher. It co-leads the international "10 Point Plan for Financing Biodiversity," sharing its playbook with other nations rich in nature but burdened by debt. It's attracting major investments from bodies like the International Finance Corporation to fund banks that lend specifically to biodiversity-friendly businesses. And it's racing toward the global "30x30" goal, having already committed to protecting 30% of its invaluable marine territory by 2030.

This is the beautiful, maddening, and inspiring paradox of Ecuador. It is a nation that struggles to follow its own best advice, yet offers the world a blueprint for a better future. The fight for Yasuní shows how hard it is to break old habits. The triumph of the Galápagos deal shows what's possible when creativity is unleashed. Ecuador's journey is a messy, complicated, and deeply human dance between the ideal and the real. And it offers a profound lesson for us all: saving the world requires not just a change of heart, but a radical change of mind.

United States - The Hidden Costs of Abundance: Industrial Agriculture, Habitat Fragmentation, and the Endangered Species Act

America carries two faces. One is the stoic, majestic face of Yosemite's granite cliffs and the Grand Canyon's sweeping grandeur, a face we put on postcards and in our national myths. It's the America of sweeping conservation laws and protected wilderness, a global leader with a green conscience. But there is another face, one we prefer not to see. This one is scarred by consumption, etched with the lines of suburban sprawl and industrial farms. This is the face of a nation whose appetite for more is so vast that its ecological footprint stomps not only on its own backyard but across the entire planet, leaving a trail of silent extinctions in its wake. This is the story of that other face, of the deep wounds beneath the beautiful veneer.

The Great Unraveling: A Land of Lonely Islands

Think of America's wilderness not as a solid continent, but as a magnificent, ancient tapestry. Now, picture us, day by day, pulling at its threads. We don't always burn the whole cloth; more often, we just snip it into pieces. A highway here, a housing development there, a pipeline snaking across a prairie. Each cut frays the whole, turning a single, vibrant ecosystem into a collection of lonely, disconnected islands. This is habitat fragmentation, the quiet, creeping erosion of the wild.

For the creatures caught in these shrinking pockets, the world becomes a prison. Imagine being a Florida panther, a ghost of the southern woods, needing vast territory to roam and find a mate. Suddenly, your kingdom is fractured by asphalt rivers you dare not cross. Trapped, your family tree withers, and the genes passed down through millennia begin to turn on themselves, manifesting as weak hearts and fragile futures. The tapestry unravels.

It isn't just the large predators who suffer. The tiny mechanics of life break down. A bee, searching for nectar, finds its floral highway abruptly ends at a manicured lawn. A seed, meant to be carried for miles in the gut of a bear, falls uselessly at the edge of a new subdivision. The constant hum of traffic, the glare of city lights, the chemical haze—it's a relentless, low-grade warfare on the senses of every living thing that isn't human. In Nebraska, the sea of tallgrass prairie that once defined a continent is 98% gone, plowed under for our needs, leaving its wild inhabitants with nowhere to go. We are building a nation of isolated cages.

The Silent Harvest: The True Cost of Our Daily Bread

There is a terrible tidiness to the American heartland. Mile after mile of corn, or soy, or wheat, marching in perfect, sterile rows to the horizon. This is industrial agriculture, a system that feeds millions but starves the earth itself. Where a symphony of diverse grasses, flowers, and insects once thrived, there is now a single, silent note.

This landscape, doused in chemical cocktails, has become a death trap for some of our most beloved creatures. The monarch butterfly, whose epic migration is a multi-generational miracle, has seen its population crash by 70%. Its crime? To depend on the humble milkweed, a plant systematically eradicated by herbicides to make way for crops.

The poisons don't stay on the fields. They bleed into our waterways, a slow, steady transfusion of nitrogen and phosphorus from fertilizers. This chemical feast flows down the Mississippi River, where it nourishes not life, but a monstrous explosion of algae in the Gulf of Mexico. As this green sludge dies and decays, it suffocates the ocean, creating a vast "dead zone" where fish and shrimp can no longer breathe. The Gulf is, quite literally, holding its breath, choking on the runoff from our bountiful, toxic harvest. We have transformed our fields from places of nourishment into sources of blight, degrading the very soil that is the foundation of life itself.

The Uninvited: An Invasion in Plain Sight

Some threats don't bulldoze their way in; they arrive silently, hidden in the ballast water of a ship or in a pot of exotic garden flowers. These are the invasive species, the uninvited guests who have decided to stay, and they are re-engineering our native ecosystems with terrifying efficiency. They are a factor in the peril of an astonishing 42% of America's threatened or endangered species.

In the Great Lakes, the tiny, razor-sharp zebra mussel, a stowaway from Eurasia, has carpet-bombed the lakebeds, wiping out native mussels and rewriting the aquatic food web. In the East, a nearly invisible insect, the hemlock woolly adelgid, is drinking the life out of entire forests, felling ancient hemlock trees that have anchored our riverbanks for centuries. Their death is a quiet catastrophe, a slow-motion collapse that echoes through the whole ecosystem.

And then there is the horror-movie reality of the Florida Everglades, where the Burmese python, a creature of Southeast Asia born of the exotic pet trade, has become the apex predator. These colossal snakes are swallowing the Everglades whole—from raccoons and birds to full-grown deer and alligators. They are a shadow that has fallen over the "river of grass," consuming its life force and leaving a profound silence in their wake. These invaders thrive in the chaos we've created, turning our disrupted landscapes into their empires.

The Fever: A World Under Stress

Beneath all these pressures, a global fever is rising. Climate change is no longer a future forecast; it is the heat on our skin and the smoke in our lungs. It is the great amplifier, turning every other threat up to eleven.

In the American West, the fire season is no longer a season; it's a year-round state of anxiety. The land is hotter and drier, turning forests into tinderboxes. Fires now rage with a ferocity that incinerates the soil, sterilizing the ground so that the forests we knew may never grow back. In their place, a different, poorer landscape emerges, a grassland savanna where a mighty forest once stood.

The heat is stressing every system. The iconic salmon of the Pacific Northwest, creatures of cold, clear water, find their rivers warming into a bath they cannot survive. The rising oceans, swollen by melting glaciers, are relentlessly consuming our coasts. The tiny sand islands of Hawaii, where endangered monk seals have pupped their young for generations, are simply vanishing beneath the waves, erased by the rising tide. We are losing not just species, but the very ground beneath their feet.

The Ghost in Our Shopping Cart: America's Imported Extinction

Perhaps the most difficult truth to face is that our nation's conservation report card has two grades: one for what we do at home, and another, hidden one for the damage we pay others to do on our behalf. We have outsourced our destruction.

A stunning 2025 Princeton study, "Global Biodiversity Loss from Outsourced Deforestation," pulled back the curtain on this arrangement. It revealed that the coffee, chocolate, timber, and palm oil we consume in the U.S. are directly fueling the destruction of forests in the world's most vulnerable and vital hotspots, particularly in Central America and Southeast Asia. The study acted like a forensic accountant for nature, tracing the supply chains from a cleared patch of rainforest to the products in our hands. The findings were a gut punch: the biodiversity loss caused by the consumption in wealthy nations like the U.S. was, on average, 15 times greater abroad than within their own borders.

Every time we make a purchase, we are casting a vote for the world we want to live in. Too often, without knowing it, we are voting for the extinction of a species thousands of miles away. The true ecological footprint of the United States isn't just the land within its borders; it's a shadow that stretches across the globe.

A Chorus of Rebels: Hope in a Fractured Nation

In the face of this daunting picture, a new kind of American leadership is emerging, not from a unified federal government, but from a patchwork of determined states and cities. While national policy often feels paralyzed, a rebellion of hope is taking root at the local level.

California is forging a path, pledging to protect 30% of its lands and waters and backing it with billions of dollars for wildlife corridors and restoration projects led by Indigenous communities, honoring the wisdom of the land's original stewards. On the Klamath River, Oregon is undertaking the largest dam removal project in history, a monumental act of healing designed to liberate a river and welcome the salmon home. Even in staunchly conservative Wyoming, a deal was struck to protect state land within Grand Teton National Park forever.

These are not small gestures; they are defiant acts of creation in an age of destruction. They are proof that a different way is possible. Yet, this patchwork, for all its beauty, is not enough. A migratory bird doesn't recognize state lines, and a changing climate won't be stopped by a single city's solar panels. These islands of progress exist in a sea of national inertia.

The American paradox remains our central challenge. We are a nation that treasures its wild inheritance while simultaneously fueling its destruction. We build monuments to nature while our daily lives dismantle it. The path forward requires a new kind of vision, one that sees the connection between the coffee in our cup and the fate of a distant forest, between the highway we build and the silence of a species. It requires us to look at both of our nation's faces, the beautiful and the scarred, and finally choose which one we will show the world.

We've been keeping two sets of books for our country. One is written in the bold, black ink of progress and profit. It tells a story of booming production, of cities reaching for the sky, and fields harvested with breathtaking efficiency. The other ledger is written in invisible ink, on the wind and in the water. It's a book of debts, of silent losses and creeping voids. This is the story of what we've traded away.

The intricate web of life that holds our world together is fraying, and we are the ones holding the scissors. In the United States, the fading chorus of birdsong and the disappearance of native wildflowers aren't just a tragedy for nature lovers. They are entries in our hidden ledger, accounts coming due with staggering costs to our economy, our health, and the very stability of our future. Protecting nature isn't a gentle ideal; it's a ruthless economic necessity.

The Illusion of the Perfect Field

Picture a perfect American field. It rolls on for miles, a seamless, emerald carpet of corn or soy, swaying under a vast sky. It's a triumph of engineering, a food factory without a roof. But stand at its edge and listen. The silence is the first clue that something is wrong. Where are the buzzing bees, the meadowlarks, the rustle of small creatures in the undergrowth? This isn't a landscape; it's a biological desert, painted green.

This is the unnerving heart of industrial agriculture, the single greatest driver of our nation's biodiversity loss. In our quest for maximum yield, we've waged a quiet war on diversity itself. We've bulldozed the messy, thriving chaos of grasslands and forests, replacing them with sterile monocultures. This efficiency has a brutal cost: agriculture threatens over 86% of America's at-risk species, leaving them with nowhere to live, eat, or raise their young.

Into these silent fields, we pump a chemical cocktail. Fertilizers, gorged with nitrogen, spill from the soil with every rain, a toxic flush that poisons our rivers. This runoff is a death sentence for our waters, sparking monstrous algae blooms that choke the life from everything below. It's why a "dead zone" the size of a small state materializes in the

Gulf of Mexico each year—a testament to the unseen consequences of the food on our plates.

Then come the pesticides, the blunt instruments we use to kill our enemies. But they can't tell a pest from a pollinator. They are a poison that doesn't discriminate, killing the very bees and butterflies that gift us with apples, almonds, and countless other foods. This isn't just an ecological loss; it's a direct threat to our food supply. And as these chemicals accumulate in the soil, they kill the vibrant underground world of microbes and fungi, turning rich earth into lifeless dirt, creating a vicious cycle where we need more chemicals just to grow anything at all.

This system finds its apotheosis in Concentrated Animal Feeding Operations (CAFOs)—vast, crowded cities of livestock. The rivers of waste they produce foul our water and air, while their insatiable demand for feed drives the relentless expansion of the very monocultures that started this cascade of destruction.

The Concrete Fracture

Beyond the farm, our cities and roads spread like a crack across a pane of glass. We see it as growth, but for the wild world, it is a violent fragmentation. A new highway that slices through a forest is more than just a road; it's a wall. It isolates populations of animals, turning vast, interconnected territories into small, desperate islands of nature.

For a bobcat or a coyote, this means a shrinking world, a deadly game of Frogger just to find a mate or a meal. For entire populations, this isolation is a genetic dead end. Trapped in their green prisons, they lose the diversity that is their best defense against disease and a changing climate, inching them closer to extinction.

Our urban world brings other pressures. The concrete and asphalt create "heat islands" that bake sensitive species. The perpetual twilight of light pollution disorients migratory birds, luring them into fatal collisions. The constant drone of traffic masks the calls of mates and the warnings of predators. We have built our world so loudly that nature can no longer hear itself think.

A Ghost Story: The Bat and the Billions

The costs in our hidden ledger are not abstract. Let me tell you a true ghost story. It's about the catastrophic decline of America's bats, felled by a ghostly white fungus that has swept through their caves. Bats are nature's night shift, a silent, flying pest-control service of almost unimaginable value.

A landmark 2024 study connected the dots, and the picture it revealed is horrifying. As bat populations vanished between 2006 and 2017, farmers, deprived of their natural allies, were forced to douse their fields with 31% more pesticides. The direct cost of this chemical arms race and the crop damage bats would have prevented? A staggering $26.9 billion.

But the story gets darker. That invisible chemical rain didn't just stay in the fields. The study traced it into human communities, linking the increased pesticide exposure directly to a 7.9% rise in infant mortality in the hardest-hit counties. The social cost of these lost lives was valued at $12.4 billion. This is the true price of a broken ecosystem. It's not just about money; it's about the health of our children. The death of a single species cast a shadow that reached all the way into our cradles.

The Uninvited Guests Who Never Leave

As if our self-inflicted wounds weren't enough, we face an invasion. Non-native species, hitchhiking on our global trade routes, have stormed our shores and taken root. From the zebra mussel, whose clogging grip costs our power plants and water systems millions, to the emerald ash borer, a jewel-toned assassin that has erased tens of millions of trees from our forests and cities.

These invaders are a constant, hemorrhaging drain on our economy, costing an estimated $137 billion every single year. That's a tax of over $400 levied on every man, woman, and child in America, paid to fight a war against things like feral hogs tearing up farmland and Burmese pythons swallowing the Everglades whole.

The Work of Mending

But the hidden ledger holds one more entry: a story of hope. It's the story of ecological restoration, an economy of healing that is quietly booming. This isn't just about planting trees; it's about rebuilding the living systems that sustain us. And it's creating real, tangible wealth.

A 2015 study found that the work of mending nature already generated $25 billion in economic output and supported over 221,000 jobs. These are the jobs of hydrologists designing new wetlands, of skilled workers reintroducing native plants, of engineers freeing dammed rivers. It is the work of turning a liability into an asset.

Investing in restoring a coastal wetland is vastly cheaper than building a seawall and provides the priceless bonus of clean water and revived fisheries. Investing in healthy forests is a far more effective strategy than simply fighting bigger and bigger wildfires. This isn't spending; it's the wisest investment we can make.

Balancing the Books

The two ledgers lie open before us. One book, written in the ink of short-term gain, is filled with red. Its pages tell a story of staggering debts—to our economy, to our health, to our future. The other book tells a different story. It's a story of restoration, of resilience, of an economy built not on extraction, but on healing.

We are the bookkeepers of this planet. The choices we make—the food we buy, the policies we support, the way we build our communities—are the entries we make each day. We can continue writing in red ink, pretending the debts will never be called in. Or we can pick up a new pen and begin to write our way back to life, filling the pages with the green ink of a world mended, a future restored, and a prosperity that is shared by all of creation. The final accounting is up to us. Thank you.

A Nation's Wild Heartbeat

In the sprawling, beautifully chaotic expanse of the United States, there's no single blueprint for saving a species or a forest. This isn't a country that moves in lockstep. Instead, protecting our wild heritage is a story told in a thousand different voices—a rancher in Montana, a scientist in Florida, a tribal leader in the Southwest, a city volunteer in Idaho. It's a complex, sometimes clashing, symphony of laws, local efforts, and deeply held beliefs about land and liberty. This is the story of how America, in its own stubborn and inventive way, fights for its wild soul.

The Endangered Species Act: A 50-Year Promise

Imagine a safety net, woven over half a century, strong enough to catch 99% of the species that fall into it. That is the Endangered Species Act (ESA). Born in 1973, it's a radical promise: we will not let our fellow creatures vanish without a fight.

And what a fight it has been. We've all seen the soaring symbols of its success. The bald eagle, pulled back from the brink to grace our skies once more. The gray wolf, whose howl is returning to ancient landscapes. The American alligator, a reptilian giant that now thrives in sun-drenched waters.

But the ESA's greatest triumphs are often the quiet ones. It's in the Chiricahua leopard frog, now splashing in newly restored ponds. It's in the seabeach amaranth, a tiny, tenacious plant stitching our dunes back together after a century of absence. This Act is the legal muscle behind these comebacks, working through a few powerful ideas:

The List (Section 4): This is the emergency room triage. Based purely on science, we identify species as "endangered" or "threatened." This is where we say, "This life form is in trouble, and we must act." It also allows us to map out the "critical habitat" they need to survive.

The Guardian (Section 7): This is the Act's conscience. It requires every federal agency—before building a dam, a highway, or anything else—to stop and ask, "Will this action harm a species on the list?" It forces a conversation, a moment of pause to consider the lives in the balance.

The Shield (Section 9): This is the simple, powerful prohibition against the "take" of a listed animal. You cannot harm, harass, or kill them. It's a fundamental shield that protects the vulnerable from direct harm.

The Tug-of-War

Of course, a law this powerful doesn't exist without friction. Critics point out that very few species—less than 1%—have fully "recovered" and been taken off the list. But is that the right measuring stick? Recovery isn't a switch you flip; it's a long, slow journey back from the edge of oblivion. For many species, just halting the freefall is a monumental victory.

The deeper struggles are often painfully practical. Chronic underfunding means brilliant recovery plans can end up sitting on a shelf, starved of the dollars needed to bring them to life. Bureaucratic gridlock can leave species waiting years for protection while their numbers dwindle.

And then there's the most human challenge of all: the intersection with private land. About 75% of listed species call private property home. This can create a tense standoff between conservation goals and a landowner's livelihood. To bridge this gap, we're getting smarter. We're moving from conflict to collaboration with tools like Safe Harbor Agreements and Habitat Conservation Plans. These are handshakes, not handcuffs, offering landowners stability and support in exchange for becoming partners in stewardship. It's the slow, hard work of finding common ground.

Dancing to a Different Drummer on the World Stage

When the nations of the world gather to set global biodiversity goals, one chair is conspicuously empty. The United States, alongside the Vatican City, has never ratified the Convention on Biological Diversity (CBD). We are not a party to the big international agreements, like the recent Kunming-Montreal Global Biodiversity Framework.

Why walk such a solitary path? The reasons are woven into the American fabric: a deep-seated defense of national sovereignty, a wariness of international regulations, and fears about the economic cost. The U.S. signed the treaty in 1992, but the Senate has never given the final nod.

This means we don't have a formal vote in setting the world's conservation agenda. Yet, it doesn't mean we're idle. The U.S. simply prefers to lead through bilateral partnerships, financial aid, and scientific support, charting its own course rather than joining the global convoy.

America the Beautiful: A Promise Made at Home

So, if we aren't part of the global 30x30 plan, what are we doing? We're building our own. The "America the Beautiful" initiative is the nation's answer—a bold pledge to conserve 30% of our lands and waters by 2030.

But this is 30x30 with a distinctly American flavor. It's not about top-down government mandates. Its heart and soul are voluntary, locally-led, and incentive-based. It's built on a radical idea: that the best conservation comes from the ground up.

This initiative is a massive collaboration, bringing everyone to the table:

States and Tribal Nations, honoring their deep historical and ecological knowledge.

Private farmers, ranchers, and fishers, recognizing them as the stewards of the vast majority of our working landscapes.

Cities and local communities, empowering them to become engines of conservation.

Take Boise, Idaho, for example. The city is embracing its own 30x30 goal not with regulations, but with community spirit. They're mobilizing "Weed Warrior" volunteers to restore natural areas and running "Adopt-a-Habitat" programs. This is what "America the Beautiful" looks like in practice: local people, rolling up their sleeves for the places they love. It's about making conservation a shared project, not a bureaucratic chore.

The Laboratories of Life: State-Level Innovation

With the federal government often moving slowly, individual states have become buzzing laboratories of democracy and conservation. This has created a vibrant, if messy, patchwork of protection across the country.

States like California, Nevada, and New Mexico are leading the charge, passing their own 30x30 goals and creating powerful state-level endangered species acts. But they aren't alone.

Florida is spending billions to acquire and protect critical wetlands.

Maryland is laser-focused on restoring the Chesapeake Bay.

Washington is building wildlife corridors so animals can roam safely.

New York is protecting vast tracts of wilderness in the Adirondacks.

This decentralized approach is a double-edged sword.

The Upside? It allows for tailored solutions. What works for the Florida panther won't work for the Washington elk. It spurs creativity and allows states to respond quickly to local challenges.

The Downside? It can create a confusing jumble of rules. A species might be protected in one state but vulnerable just across the border. And states with fewer resources or less political will can become weak links in the chain of survival. It's a system defined by its dynamism and its disparities.

Following the Money: A River Fed by Many Streams

There is no single pot of gold for biodiversity in the U.S. Instead, funding flows like a great river fed by countless streams, big and small.

Federal Headwaters:
The federal government provides the largest currents. The Bipartisan Infrastructure Law recently poured $3 billion into ecological restoration. Long-standing programs from the Department of Agriculture (USDA) pay farmers to create habitat on their land. And beloved mechanisms like the Land and Water Conservation Fund and the Federal Duck Stamp (funded by hunters) have protected millions of acres of irreplaceable habitat.

State & Local Tributaries:
States contribute their own powerful flows through legislative budgets, voter-approved bonds, and

dedicated funds from things like lottery tickets or hunting and fishing licenses.

Private Springs & Creeks:
Perhaps the most unique part of the American system is the enormous role of private money. Philanthropic foundations and individual donors contribute billions. And nimble, passionate land trusts work directly with landowners to protect land one acre at a time, often moving faster than any government agency ever could.

We are even exploring new financial frontiers like biodiversity credits and impact investing. Yet, unlike many other nations, we don't have a single, unified national biodiversity finance plan (like a BIOFIN). Our approach is more organic, more chaotic, and entirely our own.

A Wild, Unfinished Story

The American path to protecting biodiversity is not a straight, paved highway. It's more like a network of game trails—winding, overlapping, sometimes disappearing, but always moving forward. It's a system forged from a fierce belief in individual action, local wisdom, and voluntary partnership.

From the bedrock promise of the Endangered Species Act to the grassroots energy of "America the Beautiful," this is a story of a nation constantly reinventing what it means to coexist with the wild. It's messy, it's complicated, and it's a reflection of the country itself—a chorus of millions of voices learning, step by step, how to care for the land that defines us all. And that story is far from over. ♥

Part II: Africa & Madagascar - The Resource Curse and Endemic Fragility

Democratic Republic of Congo - Conflict, Resources, and Conservation: Protecting Great Apes in a Volatile Landscape

In the heart of Africa, there is a sound. It is not the shriek of a chainsaw or the crack of a rifle, though those noises scar the air with terrifying frequency. It is a deeper sound, a great green hum. It is the exhalation of 155 million hectares of rainforest, a living, breathing cathedral of such immense scale it dictates the weather of a continent and the health of a planet. This is the Democratic Republic of Congo, home to the Earth's second lung, a place of impossible beauty and incomprehensible sorrow.

The DRC is a ghost at the world's feast. It is a land blessed with every treasure—cobalt for our batteries, coltan for our phones, diamonds for our vanities, and a forest that scrubs the carbon from our skies. By any sane logic, its people should be stewards of a global Eden, wealthy and secure. But logic fled this place long ago. Instead, the nation's wealth is a curse that has poisoned its own soil, fueling wars that have stolen generations and nurturing a poverty so absolute it forces a mother to choose between her child and a tree.

This is not just a story about a forest. It is a story about us. It is about the cavernous distance between our glowing pledges at climate summits and the mud-and-blood reality for the people on the front lines. It is about a planetary jewel held in a hand trembling with hunger and fear. And it is about the fate of our closest relatives—the gentle Grauer's gorilla, the wise chimpanzee, and the peaceful bonobo—whose soulful eyes watch their world burn, asking a question we have yet to answer.

The Pulse of a Living World

To speak of the Congo Basin is to speak of Genesis. This is not a park; it is a planet's life-support system, the largest net carbon sink left in the world, doing more to cleanse our atmosphere than even the fabled Amazon. The DRC is the anchor of this green ocean, a nation that holds more of this treasure than any other.

To walk within it is to be humbled. Sunlight dapples through a canopy a hundred feet high, illuminating a world of staggering diversity. This is Africa's cradle of life, boasting more species of mammals and birds than any other country on the continent. It is the flash of a thousand butterfly wings, the secret rustle of the okapi in the undergrowth, the mournful call of a bird whose name is known only to the forest itself.

Here, humanity can look in a mirror. In the deep green shadows, the last communities of bonobos live out their peaceful, matriarchal lives—an entire species found nowhere else on Earth, a testament to a different path a primate society can take. Nearby, their cousins, the chimpanzees, navigate their complex political worlds. And in the eastern highlands, the Grauer's gorilla, the largest primate on Earth, a sage of the forest, moves with a gentle gravity that belies its immense power. Its population has been decimated, its survival hanging by the thinnest of threads.

These forests are also the last great stronghold of the African forest elephant, the silent gardener of this world.

Hunted to near-oblivion for ivory to fund the country's wars, these architects of the ecosystem are slowly, tentatively, returning. Their presence is a sign of health, a quiet pulse in a wounded land. The fate of this forest is their fate. And their fate is ours.

An Axe in a Mother's Hand

Forget the image of corporate bulldozers leveling the jungle. That is a monster of a different kind, one that waits menacingly on the horizon. The tragedy of deforestation in the DRC is more intimate, more desperate. It is a wound inflicted not by greed, but by need.

The primary driver, accounting for an astonishing 90% of forest loss, is the agriculture of the empty stomach. Picture a father, his muscles cording as he swings an axe. He is not a villain. He is a husband and a provider with no land, no job, no safety net. The forest is his only option. He clears a small patch, the ancient trees crashing down with a groan. He burns the remains, a practice called "slash-and-burn," which gives the thin soil a fleeting kiss of nutrients. On this patch, his wife will plant cassava and maize to feed their children for a year, maybe two. When the soil is exhausted, they will be forced to move deeper into the forest and cut again. This is not a choice; it is a calculation of survival, a slow-motion deforestation driven by the simple, gnawing pain of hunger.

The second driver is the smoke you see rising from every village, every town, every sprawling city slum. It is the bitter perfume of makala, or charcoal. In a nation where 95% of the population has no access to electricity, wood is not a lifestyle choice; it is life itself. It is the only way to cook a meal, the only way to boil water and keep your children safe from cholera. This relentless hunger for fuel has created ghostly rings of degradation around every human settlement, circles of barren earth that expand year by year. An entire economy, desperate and informal, has grown around it. Men venture deeper into the forest to bake wood in earthen kilns. They emerge, coated in soot, to transport the heavy sacks on the backs of overloaded bicycles or on the iconic wooden scooters known as chukudus, their bodies straining to earn a few francs to feed their own families.

When the Forest Becomes a Bunker

Into this landscape of daily struggle, pour the gasoline of war. For decades, the cancer of conflict has metastasized throughout the eastern DRC. The forest, once a source of life and spirit, has become two things: a bunker and a bank.

For millions fleeing violence, the deep woods are the only refuge. Look no further than the edges of Virunga and Kahuzi-Biega National Parks, UNESCO World Heritage Sites meant to be sacrosanct sanctuaries for the last mountain and Grauer's gorillas. When the M23 rebels advance, hundreds of thousands of people run for their lives, ending up in makeshift camps inside the park's borders. For a family that has lost everything, the park's ancient trees are not a global treasure; they are lumber for a shelter against the rain. The park's protected soil is not a carbon sink; it is the only place to plant an emergency crop. A park animal is not an endangered species; it is protein for a starving child. Conservation laws become a cruel joke in the face of death.

For the armed groups, the chaos is a business plan. The forest is a green ATM. They control and tax the makala trade, turning protected woodlands into charcoal factories to fund their war machine. They run illegal logging operations, smuggling ancient hardwoods across the border. They orchestrate the poaching of elephants and other wildlife. The conflict provides the perfect smokescreen for this plunder, and the plunder provides the money for the bullets that perpetuate the conflict. It is a perfect, devastating cycle, a war economy fueled by the liquidation of a world wonder.

The Guardians of a Wounded Land

To be a park ranger in the DRC is to have one of the most dangerous jobs on Earth. They are not just ecologists; they are soldiers in a war for nature itself. They are the guardians of this wounded land. Outgunned and outnumbered, they stand against militias armed with machine guns and a desperation born of poverty. Hundreds have paid the ultimate price, murdered while protecting the gorillas, the elephants, and the forest that is their shared home.

This is the brutal reality that shatters our lofty global ambitions. An international agreement signed in a pristine European city means little to a ranger staring down the barrel of a gun, or trying to explain the concept of carbon sequestration to a refugee who hasn't eaten in three days.

Yet, even here, hope is a stubborn seed. The Congolese Institute for Nature Conservation (ICCN) and its rangers perform daily acts of breathtaking courage. Visionary groups like the Virunga Alliance have realized the truth: you cannot save a tree if you have not first saved the child who lives in its shadow. They work to break the cycle, building hydroelectric plants to bring clean, cheap power to communities, weaning them off the need for charcoal. They help develop sustainable farms and create jobs that give local people a reason to protect the park rather than exploit it. It is a painstaking, fragile process, constantly threatened by the next eruption of violence, but it is the only path that leads out of the darkness.

Our Shared Heartbeat

The fate of the Congo Basin is not a remote, academic question. It is an emergency that concerns every single person on this planet. Its health is inextricably tied to the stability of our climate. Its survival is a test of our own humanity. The DRC stands at a precipice, holding a key to our planet's survival but lacking the peace and strength to turn it.

The world cannot continue to praise the forest while ignoring the agony of the people. Sending money for conservation is not enough. We must demand and support real peace, helping to silence the guns that make true protection impossible. We must invest in the Congolese people—in their schools, their clinics, their access to energy, their right to a life of dignity. We must help them build a future where the forest is worth more alive than dead.

The story of the Congo is the story of a nation with a broken heart and an unconquerable soul. It is a story of sublime beauty and supreme suffering. To turn away is to be complicit in the destruction of a global treasure. The world's second lung is gasping for air. It asks for more than our pity. It asks for our courage. For in the quiet pulse of the Congo, we may yet hear the beating of our own shared heart.

A Land That Bleeds Treasure

Beneath the rich, wounded soil of the Democratic Republic of Congo (DRC) pulses a treasure that has become a torment. This is the land's paradox: a heart of cobalt, copper, and gold that, instead of nourishing its body, poisons its veins. Every time you hold a smartphone, every time the world celebrates a new electric car, a shadow of that progress falls across the Congo. The materials that power our clean, bright future are mined from a present steeped in a relentless cycle of violence, poverty, and ecological ruin. This is the story of the resource curse, told not in statistics, but in scars.

The Whirlpool: How Poverty Feeds the Guns

For millions of Congolese, life is a daily negotiation with the earth. Survival means carving a small plot from the forest to grow food, or felling a tree for the charcoal that will cook a meager meal. It is a fragile existence, a desperate reliance on a land that is itself gasping for breath. This raw vulnerability is not a tragic accident; it is a carefully exploited business model.

Into this desperation step the ghosts of the Congo's endless conflict. Armed militias and criminal syndicates haunt the mineral-rich eastern provinces, their power rooted not in ideology, but in the earth itself. They are the gatekeepers of the mines, the illegal loggers, the poachers. The ancient forests are torn down for timber, the wildlife is hunted for bushmeat to feed the miners, and the ground is ripped open for the minerals that will buy more bullets, more power, more suffering. The wealth of the land funds the very violence that keeps its people in chains, spinning a vicious whirlpool that pulls entire generations under.

While warlords and corrupt officials build fortunes on smuggled minerals, the nation they bleed dry crumbles. Schools are husks, clinics are empty, and roads dissolve back into the jungle. Families are scattered like seeds in the wind, fleeing violence and abandoning homes that have been in their lineage for generations. In the mines, children whose hands should hold pencils instead hold pickaxes, digging their own futures into a shallow grave for a pittance that barely keeps starvation at bay. The land is plundered, and its people are left with nothing but the dust.

The Green Lie and the Red Earth

The world demands a "green transition," and the DRC has the metals to make it happen. The global hunger for cobalt and copper has ignited a mining boom that is tearing the country apart.

On one side, massive industrial mines, often operated by global corporations, promise development but deliver devastation. They leech toxins into the watershed, leaving rivers with an unnatural metallic sheen and the soil barren. For communities living downstream, this is a slow-motion catastrophe. The water that once gave life now brings sickness. Women suffer miscarriages and give birth to children with devastating defects. Skin burns and peels from contact with the poisoned streams. The air, thick with dust laced with lead, arsenic, and uranium, fills their lungs, causing chronic respiratory illness. The land meant for farming is poisoned, the fish die off, and the twin specters of poverty and hunger grow larger.

On the other side is the desperate, chaotic world of artisanal mining. Here, men and boys, armed with little more than shovels and courage, burrow into the earth's flesh. They are unregulated, unprotected, and often enslaved by the militias that control the territory. Collapses are common, burying miners alive in the very ground that was their only hope. These makeshift mines leave behind a lunar landscape of craters and cleared forests, a testament to habitat destruction that pushes the Congo's incredible biodiversity—the forest elephants, the gorillas, the okapi—closer to extinction. In this shadow economy, the forest that is torn down for the mine is also hunted for the miner's meal, a final, brutal irony.

An Economy Built on a Dying Lung

The cost of this ecocide is not just environmental; it is a looming economic heart attack. Two-thirds of all jobs in the DRC are in agriculture and forestry, sectors that are being systematically destroyed from the ground up.

A 2023 World Bank report painted the future in the stark language of numbers. It predicted that the one-two punch of climate change and environmental destruction could cause the DRC's GDP to hemorrhage by as much as 12.9% by 2050. This isn't just a figure on a spreadsheet; it's a prophecy that foresees another 16 million people cast into the abyss of extreme poverty.

This grim forecast is the price of ignoring the Congo's true wealth. The Congo Basin is the planet's second lung, a colossal rainforest that performs a service of incalculable worth. Its ability to absorb carbon and regulate global climate is valued at an astonishing $223 billion to $398 billion annually. It is a living, breathing miracle that is being liquidated for short-term profits that enrich a few and impoverish millions. The laws against this plunder exist on paper, but in the vast, ungoverned expanses of the country, they are a hollow echo. Corruption ensures the saws and the excavators keep running, and the forest keeps falling.

The Price We Cannot Calculate

Beyond the economic models and environmental reports, the cost is measured in human souls. It is the silent terror of a family fleeing their village in the dead of night. It is the chronic ache of a child's empty stomach in a nation that consistently ranks among the most food-insecure on Earth. It is the grief of a mother watching her community sicken from water she knows is tainted, but has no choice but to use.

This is the cycle, brutal and unbroken. Poverty forces a reliance on the land. That reliance opens the door for violent exploitation. That exploitation shatters the environment, triggering health crises and economic collapse. And that collapse plunges people even deeper into the poverty that started the cycle anew.

To break it is to untangle a knot of human desperation, corporate greed, and our own global appetite. For the DRC, the path to a future of peace and prosperity is not paved with the minerals being torn from its earth, but in healing the very land that has been forced to bleed its treasure for far too long.

The Emerald Paradox: Congo's Fight for Its Soul

Listen closely. Can you hear it? Beneath the drone of the global economy, another sound persists. It is the deep, steady respiration of the Earth's second lung: the Congolese rainforest. A world of emerald canopy and shadowed earth, this is a place where life spills over in staggering abundance. It's the mist-shrouded sanctuary of the mountain gorilla, the stage for the bonobo's complex dance, and a living library of biodiversity so rich we have yet to read all its volumes.

But listen again, closer this time. Another sound cuts through the symphony of life—the sharp crack of a poacher's rifle, the relentless whine of an illegal chainsaw, the hollow promise of power whispered in a backroom deal. This is the Democratic Republic of Congo (DRC), a nation defined by this profound paradox: a land of immeasurable natural life locked in a desperate struggle with the realities of human survival and ambition.

The Ghost in the Green

To be a park ranger in the DRC is to understand courage in its rawest form. Imagine stepping into the vast, cathedral-like silence of Virunga National Park. You are not just a guardian of nature; you are a soldier on a forgotten frontline. Since 1996, more than 220 of your colleagues have become ghosts in this green expanse, their lives claimed while defending it. This isn't a park in the gentle, recreational sense. It is a territory besieged.

Here, the lines blur between conservation and conflict. Armed militias use the forest's immense cloak as a hideout, terrorizing communities and transforming a world treasure into a base of operations. The weakness in government isn't an abstract concept; it's the open gate through which chaos marches. It's the official who looks the other way, the unenforced law that exists only on paper, the vacuum where order ought to be. For a local community, this means a life lived in fear. For the forest, it means death by a thousand cuts—trees felled for charcoal, rivers choked with mining runoff, and animals hunted to the brink of extinction. The cycle is as vicious as it is tragic: instability invites exploitation, and exploitation fuels yet more instability.

A Roar in a Quiet Room

Now, picture a different jungle: the sterile, air-conditioned halls of a global summit. This was the scene at COP15, where nations gathered to draft the Kunming-Montreal Global Biodiversity Framework, a grand plan to save nature. As the final agreement was gavelled in, a single, powerful voice rose in protest. It was the DRC.

Their delegate took the floor, and the polite diplomacy of the room shattered. The financial aid promised by wealthy nations—$20 to $30 billion per year for the entire planet—was not just insufficient. It was, in their words, a "pittance."

That one word hung in the air, thick with the weight of unsaid truths. A pittance, they argued, for the "opportunity cost" of not leveling their forests for palm oil plantations or open-pit mines that could offer a desperate population a foothold out of poverty. A pittance for the service their forests perform for free: scrubbing the carbon from the air the whole world breathes, anchoring the rains that fall on neighboring farms, and holding the line against climate catastrophe.

In that moment, the DRC wasn't just a developing nation asking for aid. It was a custodian of a global treasure demanding its value be recognized. It was a raw, honest, and uncomfortable indictment of a global system that consumes the benefits of nature while hesitating to pay the price for its preservation. Though they have since begun the hard work of implementing the framework's "30x30" target, their point was made. The world was put on notice.

A Currency of Survival

How do you fund the salvation of a rainforest the size of Western Europe? The DRC knows it cannot wait for the world's conscience to catch up. So, it innovates.

This isn't just about traditional aid. It's about forging new financial lifelines. A payment of $19.47 million from the World Bank's Forest Carbon Partnership Facility wasn't a handout; it was a performance bonus, earned by successfully reducing carbon emissions in the Mai-Ndombe province. Crucially, a large share of this money is mandated to flow directly to the Indigenous and local communities—the forest's original and most effective guardians. Similarly, a $500 million pact with the Central African Forest Initiative (CAFI) represents a more sustained commitment, a down payment on a greener future.

But perhaps the most creative leap is the "Bonobo Credit." This is a revolutionary gamble. What if you could create a currency based not on gold or oil, but on the life of an endangered ape? The idea is to sell these biodiversity credits to the private sector, allowing corporations and individuals to invest directly in protecting the bonobo and its habitat. It is an attempt to write the value of life onto the world's financial ledger, to create a market for survival itself.

The Crossroads of a Nation

The path forward for the DRC is a tightrope walk over a chasm. On one side lies the urgent need for development to lift millions from poverty. On the other lies the sacred duty to protect a natural heritage that belongs not just to Congo, but to all of humanity.

The numbers—millions from the World Bank, billions from COP15—can feel abstract. But on the ground, they translate into boots for a ranger, a school for a village, or a sustainable farming cooperative that offers a family an alternative to poaching. True success, however, won't be measured in dollars alone. It will be measured in the strength of its institutions, in the silencing of the guns in its parks, and in the empowerment of its people.

The story of the DRC is a microcosm of our planet's greatest challenge. It's about finding a balance between what we need and what we must not lose. It's about recognizing that the health of a remote forest is inextricably linked to the health of our global community. The world watches, holding its breath, as the Congo fights for its emerald soul. Its struggle is not its own; it is a reflection of a global conscience learning, ever so slowly, to value the wild.

Madagascar - Endemic Species & Livelihood Pressures

The Island of Whispering Ghosts: A Requiem for Madagascar

Imagine an ark, set adrift from the rest of the world millions of years ago. On this vessel, life blossomed into forms seen nowhere else on Earth. It became a laboratory of evolution, a masterpiece of creation painted in the vibrant colors of chameleons, the deep green of ravenala palms, and the luminous eyes of lemurs. This is Madagascar. But the ark is sinking. The masterpiece is being erased.

The air, once thick with the symphony of a billion living things, now carries the scent of smoke and the sound of silence. A wound stretches across the island, a gash so vast it stains the rivers blood-red and can be seen from the cold distance of space. This is the scar of deforestation, where 80-90% of the island's original forests have vanished, not into memory, but into ash and barren earth. We are not just losing trees; we are losing entire worlds, unique chapters of Earth's story torn out before they could be fully read.

The Ghost in the Trees

To know Madagascar is to know the lemur. They are the island's soul, its most famous and fantastical children. With their spectral gaze and nimble grace, they are the primates of a parallel universe. Today, they are the most endangered mammals on the planet.

This is not a statistic; it is a tragedy unfolding in real-time. For every one hundred lemur species, ninety-eight are now whispering their farewells. Imagine the forest canopy, once a bustling metropolis of leaping bodies, now falling quiet. At least seventeen species have already become true ghosts, their calls forever silenced. The remaining cling to life in shattered fragments of forest, tiny green islands in a sea of destruction. They are being starved out, hunted, and isolated, their homes shrinking until there is nowhere left to leap. We are witnessing the dimming of the island's brightest stars.

A Fevered Land, a Thirsty Sky

A new, invisible threat now settles over the weakened land: the shadow of a changing climate. The island's ancient ecosystems, already wounded, have lost their power to heal. In the south, the earth cracks and turns to dust. The rains have forgotten their way here. This is not a future prediction; it is a present reality. The 2021 drought was not just a dry spell; it was a famine born of a globally altered climate, a humanitarian crisis where people, hollowed by hunger, were forced to consume the last of their natural inheritance to survive.

The sky weeps too much or not at all. Cyclones of unprecedented fury tear at the coasts while the interior bakes. This fever is disrupting every delicate rhythm of life, pushing both people and wildlife into a desperate, shared struggle on an increasingly inhospitable ark.

The Hunger that Fells a Forest

The crisis of Madagascar's wild places cannot be understood without looking into the eyes of its people. In one of the world's poorest nations, survival is a daily battle fought on the front lines of the natural world.

The axe that fells a rosewood tree is often swung by a father with no other way to feed his family. The fire that clears a patch of forest—a practice called tavy—is set by a farmer whose children are hungry today, not in some distant, abstract future. It's a devastating cycle: the fire enriches the soil for a season or two, but soon the earth is spent, and the rains wash its life into the sea. The farmer must then move deeper into the forest, leaving a trail of degradation behind, a captive in a system that forces him to destroy his own legacy.

And then there is the charcoal. For 99% of Malagasy households, the hearth is fueled by the island's bones. The charcoal trade is a lifeline for some and a death sentence for the spiny forests of the southwest, where endemic trees are sacrificed to cook a meal in the city. This is not a choice born of malice, but of necessity. When poverty is this deep, the long-term health of the planet is a luxury that cannot compete with the immediate pangs of hunger.

Scars of Greed and Silent Invaders

Beyond the struggle for survival, a darker, more calculated destruction is at play. Illegal networks bleed the island of its treasures. The deep red of rosewood, the living jewels of radiated tortoises, and even the gentle lemurs are smuggled from their homes, feeding a voracious international black market. This trade thrives in the shadows of corruption, leaving park rangers outgunned and justice undone.

At the same time, a silent coup is underway. Invasive species, stowaways on the currents of human travel, are waging war on the island's natives. Feral cats and rats hunt animals that never evolved to fear them. The toxic Asian common toad poisons the water. Foreign vines like Koster's curse and strawberry guava choke the life out of native flora, creating sterile green deserts where a rich tapestry once grew.

Below the soil, another kind of extraction takes its toll. The quest for nickel, cobalt, and ilmenite carves up the land, polluting the water that communities and ecosystems depend upon. The promise of wealth often leaves behind a legacy of poisoned rivers and displaced lives, another scar on the land's battered skin.

Mending the Shattered Emerald

Yet, in the midst of this requiem, a defiant chorus of hope is rising. The spirit of Madagascar, the life force that created such wonder, is not yet extinguished. It lives in the people fighting to reclaim their heritage.

The government, in a landmark achievement, has fought to have the Atsinanana Rainforests removed from UNESCO's "in Danger" list. This wasn't a miracle; it was the result of a decade of hard work—using drones to monitor fires, empowering rangers, and strengthening laws. It is proof that the tide can be turned.

It lives in the local communities who are forging a new path, one where their well-being is entwined with the health of the forest. They are planting fruit trees instead of clearing forests, using sustainable fishing nets, and building eco-lodges that turn the beauty of their island into a source of pride and income. They are proving that humanity does not have to be a destructive force.

It lives in the ambitious goal to blanket the island in green again, planting millions of native trees to stitch the fragmented forests back together. And it lives in the global partnerships—scientists, conservationists, and ordinary citizens from around the world—who refuse to let the ark sink.

The journey to heal Madagascar is long and fraught with challenges. But every tree planted, every community empowered, every species protected is an act of defiance against the silence. It is a promise to the whispering ghosts of the forest that they will not be the last. It is a fight to ensure that the island of wonders remains a living, breathing testament to the beauty of life, not a monument to its loss.

The Heartbeat of the Axe, The Silence of the Forest

Before the sun crests the highlands of Madagascar, a sound echoes through the mist—not a bird call, but the thud of an axe. It's a sound that is both a heartbeat and a death knell. It is the sound of survival. This is not a story of villains and heroes. It is a story of hunger, of smoke-filled huts, of a mother's choice between a tree standing tall and her child's empty rice bowl. This is the story of an Eden, an ark adrift in the Indian Ocean, that is being unmade, piece by piece, by the very hands that call it home.

A Pact Betrayed by Need: The Ghost of the Soil

The old ones remember a different rhythm. They speak of tavy, the slash-and-burn, as a conversation with the land. A small clearing, a fire that returned life to the earth as ash, a season of harvest, and then, the most important part: the long, patient silence. For decades, the plot would be left to the forest, a debt repaid, allowing the great green heart of the island to mend the wound. It was a sacred pact.

Today, that pact is broken. Not by malice, but by a rising tide of human need. The fallow period, once a generation, is now a gasp—a few years, sometimes less. The soil, given no time to heal, gives up its ghost. It becomes pale and weary, offering only meager handfuls of rice or maize.

And so the farmer, his face etched with the map of his struggle, must push deeper. He and his family venture onto slopes so steep they seem to claw at the sky, clearing the ancient trees that hold the island together. When the rains come—fierce, tropical downpours—they find no canopy to greet them. They strike the earth like a fist, washing away the lifeblood of the land. The red soil bleeds into the rivers, choking them, turning the water the color of a wound. This is the serpent eating its own tail: to feed his family, the farmer must destroy the very land that feeds him, pushing him deeper into a forest that grows smaller with every hungry season.

Black Diamonds of Necessity: The Smoke in Every Kitchen

Step into nearly any home in rural Madagascar, and you will breathe in the ghost of the forest. For millions, electricity is a distant rumor. Life is cooked over a fire. The warmth of a family, the light in the evening, the very act of making a meal is fueled by wood. A family might burn a hundred kilograms a month, a slow-burning sacrifice of the surrounding woodlands.

For those in the cities, the forest arrives in burlap sacks: charcoal. These are the black diamonds of necessity, born in remote clearings where men, with little other work, burn logs in earth-covered kilns. It is an unregulated, desperate industry that provides a meager income but carves vast, silent holes in the landscape. And with every meal cooked, the smoke stings the eyes and scars the lungs of women and children, a quiet health crisis woven from the same threads as the environmental one. The lack of an alternative—a simple, efficient stove, a small solar panel—is a chain that tethers millions to this cycle of destruction.

The Bleeding of the Rosewood Kings: A Shadow Trade

Deeper in the forest, a different kind of destruction unfolds, one driven not by hunger but by greed. This is the world of illegal logging, a shadow economy built on the blood of precious trees. Rosewood and ebony, with their deep, lustrous hearts, are the targets. On the other side of the world, they will become a rich man's furniture, a musician's instrument. Here, they are a curse.

The trade is a patient monster. When the government declares a ban, the timber barons do not stop. Their crews—often poor villagers paid a pittance for perilous work—cut the trees and hide the massive logs like sleeping giants in the jungle's depths. They wait. They wait for a moment of political chaos, a corrupt official's signature, a window of opportunity to open. Then, the ghosts of the forest are trucked out under the cover of darkness, legitimized with false papers, and shipped away.

National parks, meant to be the last sanctuaries, have become battlegrounds. In places like Masoala and Marojejy, the silence left behind by the loggers is heavier than any fallen tree. The profit flows to a phantom elite, leaving the local communities with a ravaged inheritance and the knowledge that their natural wonder was stolen while they watched.

Eating the Last Pages of a Story: The Hunt for Bushmeat

The forests of Madagascar are home to creatures of myth and magic—lemurs that leap through canopies like furry spirits, chameleons that carry the colors of the world on their skin. Today, they are also becoming food. As poverty deepens, the hunt for any source of protein intensifies. Bushmeat, once a small part of tradition, is now a desperate measure.

The wide, startled eyes of a lemur, an animal that exists nowhere else on this planet, are becoming a sight not for tourists, but for hunters. Each one taken is more than a loss of life; it is a loss of function. These animals are the forest's gardeners, dispersing seeds and pollinating flowers. Without them, the forest loses its ability to heal, to regenerate. It is like tearing out the last pages of a book you are still trying to read, ensuring the story can never be whole again.

A Famine Made of Dust and Ghosts: The Human Cost

In the south, the consequences have a name: the Kere. It means famine. But this is not simply a drought. It is an environmental collapse made manifest. It is the invoice for the vanished trees.

Where forests once held moisture and anchored the soil, there is now only dust. The wind whips it into blinding storms, burying villages in sand. The land, stripped of its defenses, offers nothing. Crops wither. Livestock, with ribs like xylophones, collapse. The Kere is a ghost that walks in daylight, hollowing out the faces of children and driving families from their ancestral homes.

They become climate refugees in their own land, fleeing the dust for overcrowded cities or other fragile territories. This exodus ignites conflict. Desperate men, known as dahalo, raid villages for the last remaining cattle, turning a humanitarian crisis into a war of neighbor against neighbor. The Kere is the final, brutal proof that when we wage war on the environment, we wage war on ourselves.

The economic toll is a slow strangulation. The tourists who once came to see the lemurs find fewer reasons to visit, and the local guide stands in an empty lodge, his stories echoing in a silent forest. The farmer prays for rain on soil that can no longer hold it. The nation, already poor, spends fortunes not on schools or hospitals, but on recovering from cyclones and droughts—disasters made infinitely worse by the lack of a natural, forested shield.

This is the story of Madagascar. It is a tapestry woven from threads of poverty, hunger, and corruption, and with every pulled thread, the beautiful, unique pattern unravels. To mend it requires more than planting trees. It requires rewriting the human story, offering a choice beyond the axe. It means investing in people so they can become the guardians of their own Eden, not its destroyers. For in the soil of Madagascar, the choice is no longer between tradition and progress, but between a shared future and a common grave.

The Island's Faltering Heartbeat: Madagascar's Fight for Life

Imagine an island breathing. Not just with the rhythm of the tides, but with the pulse of life so unique it feels like a separate creation. This is Madagascar, an ark adrift in the Indian Ocean, a laboratory of evolution that has crafted creatures of myth and wonder. Lemurs with eyes like polished amber, chameleons painted with impossible colors, and baobab trees that seem to hold the wisdom of the earth. But listen closely, and you can hear that the island's heartbeat is faltering. The struggle to save this treasure is not a simple story of parks and rangers; it is a deeply human drama of a nation wrestling with its own ghosts, its politics, and its poverty, all while holding the fate of evolutionary masterpieces in its hands.

A Sickness in the Veins: The Poison of Poor Governance

The body politic of Madagascar has long been wracked by fever. The nation's history is a story of recurring political seizures—coups and power shifts that leave the country disoriented and weak. In this climate of chronic instability, long-term healing for the environment is nearly impossible. Conservation strategies, carefully planned and funded, are often left orphaned by new regimes with new agendas. Imagine a doctor's treatment plan being torn up every few years; the patient can never recover. This volatility sends a chill through the international community, whose investments in this natural wealth crave the warmth of stability.

But the illness runs deeper than the fever of politics. It's in the nation's very skeleton—its laws, which appear strong on paper but are brittle in practice. The rules designed to shield forests and fend off smugglers often crumble upon contact with reality. The guardians of these natural cathedrals—the forest rangers—are often sent into a vast wilderness without the tools, transport, or training to stand against those who would plunder it.

And then there is the cancer: corruption. It is a malignancy that spreads silently, metastasizing through the government and economy. It's the official who turns a blind eye to the scream of a chainsaw felling a century-old rosewood tree. It's the phantom permit that allows a mine to bleed poison into a pristine river. Corruption doesn't just steal trees and animals; it steals the future, diverting the lifeblood of the nation into the pockets of a few and leaving the communities who depend on the land with nothing but dust and ash.

The Scars of Yesterday: A Legacy of Mistrust

You cannot understand the island's reluctance to heal without touching its old wounds. The shadow of French colonial rule is long and dark. Colonial policies treated Madagascar's forests not as a sacred inheritance but as a commodity, carving them up for timber. At the same time, they outlawed tavy, the traditional form of slash-and-burn agriculture that had sustained Malagasy families for generations.

For the people, tavy was more than a farming technique; it was a cultural rhythm, a way of life woven into the very fabric of their identity. The colonial ban was seen not as an act of conservation, but as an act of oppression. Suddenly, a practice of survival became a symbol of defiance. This historical scar still aches today. When modern conservationists arrive, even with the best intentions, they can be met with a deep-seated mistrust, seen as the inheritors of a paternalistic past. To heal the land, one must first heal this broken trust, proving that protecting nature is not about taking away, but about building a shared and sustainable future.

The Global Vow: A Promise of Renewal

Despite its internal struggles, Madagascar has raised its hand on the world stage, making a solemn vow to mend its broken ecosystems. The nation is a key signatory to the Kunming-Montreal Global Biodiversity Framework (GBF), a global pact to stanch the bleeding of our planet's biodiversity. This is more than a signature on a page; it is a promise to its own future generations.

This promise is being written into the country's very DNA through an updated National Biodiversity Strategy and

Action Plan (NBSAP). This is Madagascar's self-prescribed treatment plan, translating global ambitions into local actions. Central to this plan is the audacious "30x30 target"—a pledge to protect 30% of its land and sea by 2030. But the government understands a profound truth: a line on a map does not make a sanctuary. Protection is meaningless unless it pulses with life and purpose for the people who live there. True conservation must enrich local communities, turning them from bystanders into the fiercest guardians of their own inheritance.

This philosophy is the soul of BIODEV2030, a groundbreaking initiative seeking to weave the green thread of biodiversity into the fabric of the economy. It dares to ask: Can a shrimp farm coexist with healthy mangroves? Can a maize farmer also be a custodian of the soil? Can small-scale miners extract minerals without leaving a toxic scar on the earth? By working with these sectors, BIODEV2030 is attempting a form of holistic medicine, healing the whole body of the nation, not just its protected heart.

The Financial Lifeline: A transfusion of Hope

Policy and promises are the spirit, but funding is the lifeblood. And in Madagascar, the nation suffers from a dangerous anemia. A Biodiversity Expenditure Review (BIOFIN) revealed a stark reality: the government's own spending on biodiversity is less than 3% of the public budget. The country is kept alive by a constant transfusion from international donors, NGOs, and foundations. While this generosity is a lifeline, it leaves the nation's most precious asset dangerously dependent on the shifting priorities of foreign lands.

The diagnosis is daunting. To implement its national biodiversity plan from 2022 to 2025 alone, Madagascar faces a staggering funding gap of nearly $900 million. Without this massive infusion of capital, the nation's vows will remain whispers, its protected areas will become paper parks, and the island's unique heartbeat could fade into silence.

The Pulse of Innovation: New Ways to Heal

Faced with this chasm, Madagascar is not succumbing to despair. Instead, it is becoming a laboratory for financial innovation. At the forefront is the Madagascar Protected Areas and Biodiversity Fund (FAPBM), a conservation trust fund that acts like a financial pacemaker for the island. With a capital base of $157 million, the fund invests its capital and uses the earnings to deliver a steady, reliable rhythm of funding to protected areas, freeing them from the anxiety of short-term grants.

Now, even more radical therapies are emerging. Imagine a "Lemur bond," a financial instrument where investors on Wall Street or in London could fund the survival of a species, their returns tied to the flourishing of lemur populations. This isn't science fiction; it's a concept being actively developed with the World Bank and the Global Environment Facility (GEF), aiming to tap into the vast ocean of capital markets to save life on Earth.

Another tool is the debt-for-nature swap, a powerful alchemy that transforms the poison of foreign debt into the medicine of conservation. In these agreements, a portion of national debt is forgiven in exchange for the country investing those funds back into its own environment.

Most recently, a jolt of energy has arrived. The new Global Biodiversity Framework Fund (GBFF) has approved an $8.56 million project for Madagascar. This is not just money; it is a targeted injection of resources aimed at reviving threatened species and healing ecosystems in five critical sites—a direct antibody fighting the disease where it is most severe.

A Future Written in Green

The story of Madagascar is a testament to the intricate dance between sickness and healing, despair and hope. Governance, history, policy, and finance are not separate chapters; they are intertwined threads in a single, complex tapestry. A corrupt official can unravel the work of a million-dollar grant. A community empowered with a sustainable livelihood can become a more effective guardian than any fence.

The path forward is steep. But in the resilience of the Malagasy people, in the creativity of financial pioneers, and in the unwavering commitment of the global community, there is a powerful pulse of hope. The ultimate vision is of an island whose heartbeat is not just stable, but strong—a rhythm that echoes in the thriving of its forests, the health of its people, and the chorus of its one-of-a-kind creatures. The fate of this living ark rests not just in their hands, but in ours.

Nigeria - Pollution, Desertification & Resource Strain

The Fever Dream of a Giant: Nigeria's Environmental Soul-Searching

To understand Nigeria today is to witness a giant wrestling with itself. It is a nation whose heart beats with two conflicting rhythms: the thunderous, black pulse of oil in its southern veins, and the faint, dry whisper of sand advancing across its northern brow. Africa's most populous country, a colossus of culture and creativity, is in the grips of a profound fever. Its immense natural beauty, a birthright of global importance, is convulsing under the strain of a crisis that is not one thing, but everything, all at once.

This is a story written in paradox. It is the story of a land blessed with life-giving wetlands that have been transformed into a toxic graveyard. It is the story of once-fertile savannas surrendering to a relentless, creeping emptiness. It is the story of forests, the very lungs of the nation, being silenced, leaving behind a void where the roar of the lion and the trumpeting call of the elephant once echoed. We will not just examine these crises; we will listen to their human heartbeat. We will walk the oil-slicked shores with the fisherman whose nets pull up only ruin, stand with the farmer whose land has turned to dust, and listen for the last echoes of life in a rapidly vanishing wilderness. We will ask not just if Nigeria can change course, but if this giant can awaken from its fever dream and choose to heal.

An Open Wound: The Bleeding Inheritance of the Niger Delta

Imagine a place woven from water and life, a labyrinth of mangrove forests and winding creeks so vibrant it felt like the cradle of creation itself. This was the Niger Delta. Today, it is an open wound on the conscience of the world. For more than sixty years, the relentless pursuit of crude oil has been a transfusion in reverse, draining the lifeblood from the land while pumping a slow-acting poison in its place.

This is not the story of a single, tragic accident. It is the story of a thousand cuts, a daily bleeding. Picture pipelines, old and tired, weeping black tears into the soil. Imagine the eternal flames of gas flares, roaring day and night, monuments to waste that vomit soot and acid rain onto villages, turning rainwater into a bitter drink and the roofs of homes into charcoal canvases. This industrial violence has seeped into the very soul of the Delta. The mangroves, with their intricate roots like grasping hands, are drowning in sludge, their leaves coated in a permanent, greasy shroud. An estimated 5,644 hectares have vanished every year since 2016—a forest the size of a city, erased by our global thirst for fuel.

For the people of the Delta, this is not an environmental issue; it is the air they breathe, the water they drink, the end of a world they knew. It is the fisherman who returns to his family with empty, blackened nets. It is the farmer whose yams grow stunted and bitter in poisoned earth. It is a legacy of sickness, conflict, and a beauty that now lives only in the memories of the old. The wealth promised by the oil has flowed past them, leaving only the ruin behind.

The Thirsty Ghost: Life in the Shadow of the Advancing Sand

Travel north, and the crisis changes its face. Here, the enemy is not a toxic surplus, but a terrifying absence. In the Sahelian states, the air is thin and hot, and the silence is broken only by the whisper of sand. The Sahara Desert is no longer a distant neighbor; it has become a hungry ghost, a creeping, thirsty entity that consumes everything in its path.

This is what climate change feels like on the front lines. The sun burns hotter, the rains have become fickle ghosts, and the earth itself is coming undone. Human hands have hastened the decline, as the desperate search for firewood and grazing land has stripped the ground of its green armor, leaving the soil naked to the wind. Each year, 350,000 hectares of land—an area larger than some countries—is lost, surrendering to the desert's advance.

The most profound symbol of this loss is Lake Chad. Once a vast inland sea, a vibrant heart that nourished millions, it has withered to less than a tenth of its former self. Its disappearance is a quiet catastrophe, stranding fishermen in a sea of dust and forcing pastoralists on desperate journeys in search of water. This slow-motion disaster unmoors people from their histories, forcing families to abandon ancestral lands and join a tide of migration that strains the resources and patience of a nation. It is a crisis of hunger, of conflict born from scarcity, and of a deep, existential thirst in a land that is forgetting the taste of rain.

The Last Roar: A Requiem for the Forest and Its Kin

Somewhere between the bleeding south and the drying north, Nigeria's forests are breathing their last. More than 90% of the nation's original woodlands have vanished, cleared for farmland, felled for timber, or devoured by the relentless spread of cities. What remains are not sprawling ecosystems, but fragile islands of green in a sea of human activity. And on these islands, Nigeria's most iconic creatures are making their last stand.

The silence creeping through these forests is a deafening one. The lion, the very symbol of African power, has been reduced to a phantom. Its mighty roar, once a bass note in the symphony of the savanna, has dwindled to the whispers of a mere 50 survivors. The nation's elephants, intelligent and majestic, have been slaughtered to a remnant of perhaps a few hundred, their great size no match for the poacher's gun. And deep in the southeastern forests, the world holds its breath for the Cross River gorilla. With only 100 individuals left, these thoughtful primates, our closest kin, gaze out from the edge of extinction.

This loss is driven by a tragic cycle. As habitats shrink, a desperate human population turns to the forest for sustenance. The bushmeat trade, once a matter of subsistence, has become a commercial enterprise, a conveyor belt of extinction that supplies city markets. It is a complex tragedy of poverty, a crisis that threatens not only the animals but also human health, opening pathways for diseases to leap from the wild into our world. In its hunger, the nation is consuming its own soul.

The Choice: To Mend a Fevered Giant

A nation so afflicted cannot survive on hope alone. In the face of this cascading crisis, Nigeria is stirring. The government has drafted a blueprint for survival, the National Biodiversity Strategy and Action Plan (NBSAP). It is an ambitious prescription, a detailed map charting a path away from the abyss, speaking of protected lands, sustainable practices, and community partnership.

But a prescription is not a cure. The plan's enemies are formidable and deeply entrenched: the gnawing corruption that siphons away vital funds, the institutional weakness that leaves laws toothless, and the crushing weight of economic and social challenges that constantly push the environment to the back of the line.

Yet, even now, the giant's immune system is fighting back. A new generation of activists, scientists, and community leaders is rising. They are the antibodies, working tirelessly to re-plant mangroves, to build a Great Green Wall against the desert, and to stand guard over the last wild places. The fight for Nigeria's environment is more than a struggle over landscapes; it is a battle for the nation's identity. The question is whether these disparate acts of healing can coalesce into a national recovery, allowing this fevered giant to finally mend its wounds and stride, whole once more, into a sustainable future.

A Tale of Two Sorrows: The Wounded Soul of Nigeria's Land

In Nigeria, the earth tells a story of two deep and distinct sorrows. In the south, it weeps a black tear of oil. In the north, it exhales a slow, choking breath of dust. These are not separate tragedies; they are the twin faces of a single crisis, a quiet war being waged against the natural world that has nurtured its people for generations. And in the shadows of it all, a silent hunt is emptying the forests, one life at a time.

The Black Tear of the South

Imagine a fisherman in the Niger Delta, a man whose ancestors read the tides like a sacred text. Today, the shimmer on the water isn't from the sun; it's the slick, rainbow sheen of crude oil. The air, once thick with the scent of salt and mangrove, now carries the acrid perfume of industry. This is the daily reality where Nigeria's great wealth is born, and where its profound environmental ruin festers.

The oil that fuels the nation's economy is a poison to the Delta. It spills from corroded pipelines that snake through the land like old, decaying arteries. It bursts from faulty equipment or is unleashed by the desperate, angry hands of saboteurs. The cause matters little to the earth itself. The result is a creeping black death.

This poison attacks the very foundation of life. It smothers farmlands, turning vibrant green shoots into blackened, sterile stalks. It suffocates the intricate network of creeks and rivers, leaving a graveyard of fish floating on the surface. For the communities here, this is not an environmental issue; it is the end of a world. The fishing net, once a symbol of providence, comes up empty. The hoe, once a tool of creation, strikes only barren ground.

The statistics are cold, but the reality they represent is a fire of human suffering. A 60% drop in food security isn't just a number; it is the sound of a child crying from hunger. It is the hollowed-out look in a mother's eyes as she stares at an empty cooking pot. This hunger gives way to a devastating health crisis. Childhood malnutrition has surged by 24%, stealing the future from children before they have a chance to live it.

And then there is the slow, invisible violence of the pollution itself. The hydrocarbons and heavy metals seep into the water, the soil, the very air people breathe. They become part of the food, the drink, the body. The consequence is a litany of modern plagues: soaring rates of cancer, lungs scarred by tainted air, and kidneys failing under a toxic load. The land that once fed them is now killing them, and the wealth it produces flows past them, a distant dream. This cruel paradox—a land of riches populated by the poor—ignites a slow burn of injustice that too often erupts into conflict, continuing the cycle of devastation.

The Whispering Dust of the North

Travel north, and the black tear of oil gives way to the whispering dust of the desert. Here, the land is not being drowned; it is dying of thirst. This is desertification, a creeping sickness driven by a feverish climate and the relentless pressure of human need.

The rains come less and less, leaving the soil parched and cracked. But this natural hardship is amplified by human hands. Herds of livestock, too numerous for the fragile land, graze the vegetation down to the naked soil. Forests are felled for the firewood needed to cook a simple meal, stripping the earth of its protective skin. Farmers, driven by necessity, push into marginal lands, exhausting them in a few short seasons.

The land, stripped of its armor of trees and grass, is left defenseless. The wind and sun bake it, beat it, and carry it away. Once-productive pastures and farmlands surrender, becoming barren expanses of sand and dust.

The consequences are a mirror image of the south's despair. Wells whisper with dust instead of water. Crop failure becomes the norm, not the exception. The struggle for dwindling resources—a patch of fertile soil, a sip of clean water—pits neighbor against neighbor, fueling tensions that can shatter communities.

This desperation triggers a great exodus. A river of people flows from the dying rural lands into the cities, searching for survival. They arrive in urban centers that are already straining, swelling the ranks of the urban poor and creating new landscapes of social and economic distress. The nation pays a staggering price for this slow-motion catastrophe, losing an estimated $5.1 billion every year. It is a hemorrhage of money, potential, and human dignity.

The Silent Hunt

Beneath these two great sorrows lies a quieter, more intimate tragedy that unfolds across the entire nation: the bushmeat trade. It is a practice born of both custom and poverty, but its outcome is the hollowing out of Nigeria's wild heart.

The result is the "empty forest syndrome"—a haunting phenomenon where the trees still stand, the canopy is still green, but the life within has vanished. The vibrant symphony of the ecosystem has been silenced. Hunters, seeking protein for their families or a product to sell in the city, lay snares and traps that have decimated populations of grasscutters, antelopes, and monkeys.

This trade also feeds a dark international appetite for the rare and the endangered. It is why the pangolin, a creature of ancient wonder, is now one of the most trafficked animals on earth. It is why the shadows of elephants and other great beasts grow longer with each passing year.

But this silent hunt does more than kill animals; it opens a biological Pandora's Box. Close contact with wild animals—through hunting, butchering, and consumption—is a well-known bridge for zoonotic diseases to leap into the human world. It is a gamble with terrifying stakes. The origins of global pandemics like Ebola and HIV are traced back to just this kind of interaction. Every animal sold in a local market carries not just the weight of its own life, but the potential for a new public health crisis.

The Tangled Root

These three crises—the spill, the dust, and the hunt—are all symptoms of a deeper ailment. They are rooted in the tangled realities of poverty, explosive population growth, and a governance structure that has often failed in its role as steward of the land.

When people are desperate, they will make desperate choices. A farmer watching his land turn to dust will cut down the last tree for firewood. A fisherman watching his children go hungry will risk the dangers of illegal oil bunkering. Poverty and environmental destruction are locked in a vicious embrace: one deepens the other in a cycle that feels impossible to break.

To heal Nigeria's land is to heal its people. It requires more than just environmental policy; it demands a revolution of the heart and mind. It means creating livelihoods that don't require destroying the earth. It means fighting the poverty that drives the silent hunt and the felling of the forests. It means demanding accountability for the black tears that poison the Delta. Ultimately, it requires recognizing that the fate of Nigeria's people and the fate of its magnificent, wounded land are one and the same.

In the heart of West Africa, a nation's conscience is stirring. Nigeria, a land of sprawling savannahs, ancient rainforests, and mangrove-fringed coastlines teeming with life, is standing at a crossroads. For too long, the whisper of its wild places has been drowned out by the roar of development, the glint of oil on water, and the silent creep of the desert. But in the summer of 2025, a new promise was forged—a plan not just written on paper, but intended to be etched into the very soil of the nation. Nigeria's updated National Biodiversity Strategy and Action Plan (NBSAP) is more than a policy document; it is a declaration, a hope, and a monumental test of will.

A Promise Whispered Across the Globe

This isn't a solitary cry in the wilderness. Nigeria's new blueprint reaches out to clasp hands with the world, aligning itself with the Kunming-Montreal Global Biodiversity Framework. Think of it as joining a global chorus, where every nation sings a verse of the same urgent song: to halt the great unraveling of our planet's living tapestry. By adopting this shared language of conservation, Nigeria is stepping out of isolation. It's a vow to see its own emerald forests and winding rivers not just as national treasures, but as vital organs of a living, breathing planet. This global handshake unlocks a flow of shared wisdom, collaborative strength, and the potential for resources to mend what has been broken.

A Nation's Choir: Every Voice Must Sing

Perhaps the most revolutionary chord struck by the new plan is its call for a "whole-of-government and whole-of-society" symphony. The old way—tucking nature away in the quiet corner of an environment ministry—is being cast aside. This new vision recognizes a profound truth: the health of an ecosystem is written in the national budget, in the designs of its highways, in the curriculum of its schools, and in the choices made on its farms.

Imagine it: the Minister of Agriculture, once focused solely on crop yields, now champions farming techniques that heal the soil and protect pollinators. The Minister of Finance learns to calculate the true wealth of a wetland that purifies water and buffers storms. The Minister of Education ensures that a child's birthright includes not just literacy, but ecological literacy—an understanding of their profound connection to the natural world.

And this choir extends beyond the halls of government. It calls to the village elder who holds generations of traditional knowledge, the fisherman who reads the tides, the CEO whose supply chain can either destroy or restore, and the activist whose passion fuels the fire of change. It is a radical, beautiful, and profoundly difficult idea: that the sacred duty of stewardship belongs to everyone.

A Line Drawn in the Sand: Zero Loss

At the core of this promise lies a goal of breathtaking ambition: to achieve zero net loss of Nigeria's most precious, life-filled areas by 2030. This is not about slowing the bleeding; it is about stopping it. It is a line drawn in the rich, red earth, declaring that the nation's ecological crown jewels will not be sacrificed. To achieve this, any unavoidable harm must be balanced by rebirth elsewhere—a logged forest replanted, a degraded wetland restored, a new sanctuary born.

This vision is further sharpened by Nigeria's commitment to the global 30x30 target—a pledge to protect 30% of its land and sea. In a nation of over 200 million people, with immense pressure for housing, farming, and industry, this is a sacred tithe of the homeland, a testament that some places must remain wild and whole. Nigeria has also joined the ranks of the Global Ocean Alliance and the High Ambition Coalition for Nature and People, standing shoulder-to-shoulder with nations determined to lead this global charge. With support from vital partners like the Global Environment Facility (GEF), which helps fund grassroots projects, the machinery for change is being assembled.

The Chasm Between Promise and Practice

But a promise, no matter how beautifully articulated, is only as strong as the will to keep it. Between the lofty words of the NBSAP and the reality on the ground lies a deep and treacherous chasm, haunted by familiar ghosts.

The Hollow Echo of Law: Nigeria has laws meant to be shields for nature. But too often, they are paper tigers. Their roars of regulation are muted by weak enforcement, their claws blunted by slow justice. This creates a vacuum where illegal loggers, poachers, and polluters operate with a chilling sense of impunity, turning protective legislation into a hollow echo.

The Shadow of Corruption: A more insidious rot is corruption. It is a poison in the system that diverts funds meant for park rangers into private pockets, that transforms permits into commodities for the highest bidder, and that shields the powerful from accountability. It is the unseen hand that guides the chainsaw and blesses the toxic pipeline, making a mockery of any conservation effort.

The Test of Titans: The ultimate test of political will comes when conservation collides with powerful economic interests. Can the government truly regulate the goliaths of oil and gas, whose operations have left deep scars on the

Niger Delta? Can it dismantle the powerful, violent networks behind industrial-scale illegal logging that are felling its forests? True commitment means holding everyone, no matter how economically mighty, to the same environmental standard.

The $8 Billion Question: Funding the Future

The paradox is staggering. Nigeria's natural world provides services—clean water, fertile soil, stable climate—valued at over $8 billion a year. Yet, the investment in protecting this natural capital is tragically small. It's like owning a priceless treasure and leaving it guarded by a single, underpaid watchman.

Studies from nations with similar challenges, like the Philippines, reveal a biodiversity funding gap of around 80%. It is almost certain Nigeria faces a similar, if not greater, shortfall. This leaves the nation heavily reliant on international lifelines like the GEF, whose small grants have admirably supported over 60 local biodiversity projects. This external help is critical, but it's a bandage, not a cure. It highlights the urgent need for Nigeria to generate its own sustainable funding.

This is why Nigeria now raises its voice on the global stage, calling on developed nations to honor their promise under the Global Biodiversity Framework: to deliver $20 billion annually to developing nations by 2025. The plea is rooted in justice. Nations like Nigeria, which steward a disproportionate share of the world's biodiversity while facing immense economic hurdles, cannot and should not bear the cost alone. Protecting the Amazon, the Congo Basin, or the forests of Nigeria is a gift to the entire world, and the world has a shared responsibility to help pay for it.

The road ahead is not a gentle slope; it is a steep, arduous climb. Success demands more than a plan. It demands a revolution in governance, an unwavering war on corruption, and the courage to make powerful interests bend to the common good. Nigeria has written a new story for itself. The world is watching. More importantly, the children of Nigeria are waiting to inherit either a legacy of loss or a landscape of life.

Part III: Asia-Pacific - The Epicenter of Growth and Loss

Indonesia - Commodity-Driven Deforestation: The Palm Oil, Pulp, and Nickel Dilemma

To stand on Indonesian soil is to feel the planet's pulse. It's a rhythm beaten out across 17,000 emerald islands, a heartbeat felt in the rustle of leaves in forests that breathe for the world, and in the silent, swirling dance of a million fish through the planet's most vibrant coral reefs. This is a land stitched together with wonder, one of only seventeen "megadiverse" nations, a living library of creatures found nowhere else on Earth.

Below the waves lies a world that beggars belief—the Coral Triangle, the veritable "Amazon of the Seas." This is the cradle of marine creation, holding a staggering 18% of the world's coral reefs. It is the ocean's nursery, its masterpiece. Onshore, the last great forests of Sumatra and Borneo serve as the final, sacred refuges for souls as majestic and fragile as the Sumatran tiger, the gentle, intelligent orangutan, and the impossibly rare Sumatran rhino. They are the living jewels of this archipelago.

But listen closely. Beneath the symphony of life, another sound grows louder. It is the roar of the chainsaw, the groan of the earthmover. It is the sound of a treasure being sold, piece by precious piece.

This isn't a single act of destruction. It's a ghost story where the monster keeps changing its face. For years, its name was Palm Oil, a phantom ingredient in our snacks and soaps, demanding a tribute of annihilated rainforests. Then it was Pulp and Paper, its hunger for packaging and notebooks leaving behind a silent, monoculture grave where a symphony once stood. Today, the monster wears a "green" mask. Its name is Nickel, the metal messiah for our electric car batteries. In a cruel twist of irony, our quest to save the planet in one place fuels the destruction of another, carving up the earth for a cleaner tomorrow that may never come for Indonesia's forests.

This relentless cycle, this "serial boom," has left scars you can feel in the air and see from space. The most tragic are carved into the peatlands. These are not just swampy grounds; they are ancient, water-logged archives of carbon. When drained for plantations, they become tinderboxes. The fires that follow are not fires of wood, but of earth itself. They smolder for months, a subterranean inferno that exhumes millennia of stored carbon, releasing it in a toxic, choking haze that ghosts across Southeast Asia, stealing the breath from millions.

The fallout bleeds into the sea. The vibrant underwater cities of the Coral Triangle are being dynamited into rubble by reckless fishing. A slurry of pollution from coastal mines and plantations smothers the corals, turning their brilliant colors into bleached-white tombstones. Along the shores, the mangrove forests—the coast's gnarled, steadfast guardians—are bulldozed for shrimp farms. With them goes the first line of defense against the ocean's fury, and the cradle for the fish that feed the coastal villages.

This is not a victimless crime. For the Indigenous peoples of the forest, this is a soul-wound. Their ancestral lands are not just property; they are the source of their stories, their medicine, their identity. When the forest falls, a culture is erased with it. For the fisherman casting his net into a murky, lifeless sea, and the farmer watching his soil wash away in the floods that follow deforestation, this "development" is a slow-motion disaster. The profits flow to distant boardrooms, while the poison seeps into the local well.

Indonesia's leaders are caught in a paradox, making vows of conservation on the world stage while drawing up blueprints for extraction at home. They speak of a monumental 90% drop in deforestation, a claim that sounds like a miracle. Yet independent eyes, reading the satellite data, see a different truth—a reduction, yes, but closer to 50–69%. More chillingly, they see the trend reversing. They see the saws spinning up once more.

Now, as we look to 2025, the government has unveiled a plan that feels less like a policy and more like a eulogy. They intend to clear another 20 million hectares of forest for so-called "food and energy estates." It's a sanitized name for an act of immense sacrifice, a final surrender of irreplaceable, living wonder for industrial-scale agriculture.

What happens when the last tiger's roar fades into silence? What stories will be told when the coral gardens are barren and the great apes are only a memory? Indonesia is at a crossroads, but it is not walking it alone. Its struggle is humanity's struggle. The future of this archipelago, this living, breathing heart of our world, hangs by the finest of threads, and we are all holding the line.

There is a story being written across the islands of Indonesia, not in ink, but in scars. It's a story of a land blessed with an impossible green, a breathing, living tapestry of forests and seas, being consumed, chapter by devastating chapter. This isn't a single tragedy, but a relentless cycle, a relay race of ruin we call "serial exploitation." One hungry industry tears through the land,

and just as the dust settles, another arrives to feast on what remains, leaving behind a silence where life once roared.

The Ghosts of the Green Desert

For decades, the ghosts wore two faces: palm oil and pulpwood. Picture the soul of a rainforest—a chaotic, vibrant symphony of life—being bulldozed into a silent, green desert. This was the march of the plantations. An endless, uniform army of palm and acacia trees replaced the rich biodiversity, their expansion fueled by our insatiable global appetite for everything from cookies to copy paper.

Once, in 1900, a staggering 84% of Indonesia was cloaked in forest. By the new millennium, that cloak was in tatters, shrinking to less than 100 million hectares. The pace became a fever pitch. By 2012, the world watched as Indonesia outpaced even Brazil, its chainsaws screaming louder than any other nation on Earth. For years, the villain of the story was palm oil, accounting for nearly a quarter (23%) of all deforestation between 2001 and 2016.

But the story shifted in 2024. A new protagonist has taken the stage in this tragedy: the pulpwood industry. With colossal new mills rising like monuments to consumption, their demand for wood fiber is voracious. They are now the single greatest force clearing Indonesia's forests, especially in the wild heart of Borneo, known as Kalimantan. These companies devour forests within their own concessions, a loophole that mocks moratoriums and sustainability pledges. They drain the life from carbon-rich peatlands, releasing ancient carbon into the atmosphere like malevolent spirits. The smoke from their fires doesn't just cloud the sky; it chokes the planet.

The Green Paradox: A Gleaming Monster

The beast of exploitation is a shapeshifter. Its new face gleams with the promise of a green future, polished and silent: the electric vehicle. To power our clean energy transition, the world needs batteries. To build those batteries, the world needs nickel. And Indonesia, holding the planet's largest nickel reserves, has become the new frontier.

This is the planet's most bitter paradox. The silent, clean electric car gliding through a city in Europe or America owes its existence to the violent gouging of the earth in a place like West Papua. Open-pit mines, like raw, red wounds on the skin of the planet, have already erased over 193,000 hectares of forest between 2000 and 2023.

The damage doesn't stop at the treeline. It bleeds into the sea. In the Raja Ampat archipelago, the "Amazon of the Seas," a realm of staggering marine beauty holding 75% of the world's known coral species, a slow death has begun. Runoff from the mines washes a suffocating blanket of sediment over the reefs, choking the coral and turning the once-turquoise water a murky brown. Indigenous communities, whose lives are woven into the health of these waters, now watch their livelihoods and traditions dissolve. They speak of children afraid to swim, of skin diseases from the poisoned sea. We are trading one of the planet's last pristine Edens for a solution to the pollution we created elsewhere.

The Scars We Live With

The consequences of this relentless plunder are not abstract statistics; they are lived realities, etched into the health, economy, and even the very geography of the nation.

A Season of Smoke: The most visceral scar is the haze. The fires, often deliberately set to clear land, burn deep into the peat, becoming nearly inextinguishable. They unleash a choking shroud that blots out the sun for months. The 2015 fires were apocalyptic, blamed for the premature deaths of up to 100,000 people across the region. The economic cost is staggering—a $16 billion blow in 2015, and another $5.2 billion in 2019, equivalent to 0.5% of the nation's GDP. It's a price tag written in scorched earth and stolen breath.

A Sinking City, A Fleeing Capital: In a move that speaks volumes, Indonesia is abandoning its own capital. Jakarta, a megacity of 10 million, is a city devouring itself. It is sinking into the sea—some parts by an astonishing 14 feet—as its residents, desperate for clean water, illegally drain the aquifers beneath their feet. Suffocating in pollution and choked by traffic, Jakarta has become a symbol of environmental collapse.

The solution? A monumental gamble: to build a new capital, Nusantara, a "sustainable forest city," from scratch in the forests of Borneo. But critics fear this isn't an escape, but an exportation of the problem. Building a city threatens orangutan habitats, displaces Indigenous peoples, and puts immense pressure on Borneo's remaining ecosystems. Is this a bold solution, or simply the next chapter in the same destructive story, leaving the poor to drown in the ruins of the old world while the wealthy flee to the new?

The Silent Seaweed: Sometimes, the greatest tragedies are the quietest. On the small island of Nusa Lembongan, the rhythm of life was once tied to the sea. For 30 years, families cultivated seaweed, a gentle dance between the moon and the tides that provided a stable life. But the dance has stopped. A perfect storm of unchecked tourism development, which polluted the waters and destroyed protective mangroves, combined with the warming, acidifying touch of climate change, killed the crops. The industry collapsed. The seaweed died, and with it, a way of life, leaving behind a silence in the water and a hole in the community.

A Future Haunted by the Past

Indonesia stands at a crossroads, haunted by the ghosts of the forests it has lost and threatened by the gleaming new monsters of global demand. The relay race of ruin—from timber to palm oil, from pulpwood to the nickel in our "clean" machines—reveals a fundamental, global truth: our consumption has consequences that ripple across the planet.

Without a profound shift toward true sustainability, responsible governance, and a global conscience, the story of Indonesia will continue to be written in scars. Every battery will hold a ghost. Every clean machine will cast a long, dark shadow over a land that is running out of time.

In the heart of the Coral Triangle, where volcanoes touch the clouds and a symphony of life echoes from rainforest canopy to ocean floor, lies a nation at war with itself. Indonesia, a paradise of unparalleled biodiversity, is caught in a great and difficult struggle. It is a tug-of-war between two powerful desires: the urgent pull of economic progress and the sacred duty to protect the natural wonderland it calls home. This is the story of Indonesia's emerald paradox, a landscape where policies of preservation and plans for exploitation grow side-by-side, like strangler figs on a sacred banyan tree.

A House Divided: The Schizophrenia of Governance

To look at Indonesia's policies is to witness a national split personality. With one hand, the government offers gestures of profound environmental stewardship. It once laid down a moratorium on new palm oil permits, a bold attempt to slam the brakes on an industry whose relentless expansion had consumed vast tracts of ancient forest. It has meticulously crafted a new Biodiversity Strategy and Action Plan (IBSAP) for 2025-2045, a visionary document that sings in harmony with the global Kunming-Montreal Framework, promising to safeguard the nation's natural heritage for generations to come.

But with the other hand, it signs off on development that feels like a betrayal of those very promises. This is where the deep, unsettling contradictions appear.

Consider the government's own Forestry and Other Land Use (FOLU) Net Sink 2030 plan. The title promises a climate triumph: a future where Indonesia's forests absorb more carbon than the nation emits. Yet, buried within this green-sounding plan is a shocking allowance: a deforestation quota of 10.43 million hectares. It is a permission slip to clear an area nearly the size of Sumatra's remaining forests. Environmental watchdogs like Greenpeace were aghast, questioning how a nation can promise to be a carbon "sink" while simultaneously sanctioning the destruction of the very forests that make it possible. It's like trying to fill a bathtub while drilling a massive hole in the bottom.

This dissonance echoes in the nation's energy sector. While the world looks toward a renewable future, loopholes in Indonesian policy allow for the quiet construction of "captive" coal plants. These are private power stations, often built to fuel industrial smelters or factories, that operate in the shadows—exempt from emissions trading schemes and often free from reporting requirements. They are a ghost fleet of polluters, tethered to the past, undermining the country's public climate commitments with every puff of smoke.

This conflict was cemented into law with the controversial Omnibus Law on Job Creation. Designed to slash red tape and attract investment, critics decry it as a wrecking ball to decades of environmental protection. It weakened the mandate for environmental impact assessments, silenced public opposition, and controversially removed the "strict liability" principle that once made it easier to hold companies accountable for forest fires. In the name of growth, the law seemed to prioritize the bulldozer over the biosphere, tilting the scales decisively in favor of exploitation.

A Promise to a Planet in Peril

Despite its internal conflicts, Indonesia looks outward with ambitious promises. As a signatory to the Kunming-Montreal Global Biodiversity Framework, it has embraced one of the planet's most audacious conservation goals: the "30x30" target, a pledge to protect at least 30% of the world's land and sea by 2030.

For the world's largest archipelago, this is a monumental task. Indonesia has set a national goal to protect 30% of its vast marine territory—a liquid kingdom teeming with more marine biodiversity than anywhere else on Earth. Yet, it has given itself a longer runway, aiming for the full 30% by 2045, with a milestone of 10% by 2030. This timeline speaks volumes about the staggering challenges. How do you police a protected area scattered across 17,000 islands? How do you balance the survival of a coral reef with the needs of a fishing village whose ancestors have harvested these waters for centuries? Success hinges not just on government decrees, but on winning the hearts and minds of coastal communities, weaving their traditional wisdom into modern conservation.

The Currency of Conservation: In Search of Green Trillions

Protecting paradise isn't free. A landmark study by the Biodiversity Finance Initiative (BIOFIN) revealed a cavernous gap between the money Indonesia has and the money it needs to fund its conservation dreams. The state budget and funds from global partners like the Global Environment Facility (GEF) are critical, but they are just drops in a bucket.

To bridge this chasm, Indonesia has become a global pioneer in a surprising field: innovative Islamic finance. Its most celebrated creation is the Green Sukuk, or "Islamic green bond." Unlike a conventional bond, which is a loan, a sukuk grants the investor a share of ownership in a tangible asset, making it compliant with Islamic law (Sharia). By issuing Green Sukuk, Indonesia has brilliantly tapped into a global pool of ethical capital, raising billions of dollars for projects earmarked as sustainable. While much of this has funded vital water management and transport infrastructure rather than pure renewable energy, it represents a revolutionary fusion of faith and finance—a model for the world.

The innovation doesn't stop there. Indonesia is exploring:

Ecological Fiscal Transfers (EFTs): A system that financially rewards regional governments for being good stewards of nature. If a province protects its forests, it gets more funding from the central government. It's a simple, powerful idea: make conservation pay.

Faith-Based Funds: Mobilizing Zakat (obligatory charity) and Waqf (charitable endowments) for green causes. With one of the world's largest Muslim populations, the potential is immense. Imagine transforming these traditional streams of generosity into a river of funding for community-owned solar panels, eco-friendly farming, and forest restoration.

The Paper Promises of the Market

Yet, even as these green shoots of financial innovation appear, a stubborn question remains: Can we trust corporate promises? For years, the world's largest palm oil and pulp companies have made sweeping "No Deforestation, No Peat, No Exploitation" (NDPE) commitments. These were meant to be market-based solutions, assuring consumers that their products weren't stained by environmental destruction or human rights abuses.

The reality, however, has been profoundly disappointing. Report after report reveals that deforestation and exploitation persist deep within the supply chains of these very companies. The promises, it seems, were written in disappearing ink. Amnesty International has documented horrific labor abuses on plantations supplying global brands, including child labor and exposure to toxic chemicals. This gap between corporate pledges and on-the-ground reality is a stark reminder that voluntary commitments are no substitute for rigorous, independent verification and real accountability.

The Path from the Crossroads

Indonesia stands at a momentous crossroads. It holds the future of some of the world's most vital ecosystems in its hands—the last Sumatran orangutans, the vibrant coral reefs, the carbon-rich peatlands. Its journey is a microcosm of the global challenge: a test of whether humanity can reconcile its endless appetite for growth with the finite capacity of the planet.

The nation's story is not yet fully written. It is a tale of brilliant innovation and frustrating contradiction, of bold promises and heartbreaking failures. The path forward demands a unified government, one where the ministry of environment and the ministry of economy walk in step. It requires turning corporate pledges into unbreakable laws and financial innovations into a tidal wave of change. The world watches, holding its breath, hoping that this paradoxical nation chooses to heal its divided soul and save the paradise it was entrusted to protect.

Malaysia - Palm Oil Expansion & Forest Management

The Breath of a Million Lives: A Deeper Look at Malaysia's Wild Heart

To speak of Malaysia's biodiversity is to speak of a miracle. It's easy to name the stars of the show—the soulful-eyed orangutan, the ghost-like tiger, the gentle elephant. They are the breathtaking headliners, the ones whose names we all know. But to truly understand this nation's natural wealth, you must listen closer. You must tune your ear to the hum of life that exists beneath the canopy, a symphony so vast and complex it defies simple description.

Imagine standing in the heart of a rainforest that has breathed for millennia. The air is thick with the scent of damp earth and blooming secrets. This is not just a collection of trees; it's a living library. On its shelves are an estimated 15,000 different kinds of vascular plants, from colossal trees that pierce the sky to orchids so tiny they seem like dewdrops clinging to a branch. This is the stage for over 300 species of mammals. For every famous tiger, there is an elusive clouded leopard melting through the shadows, a sun bear feasting on honey, its golden crescent a mark of the moon in the deep forest.

Look up, and the air is alive with the music of over 730 species of birds, a riot of color and sound. Look down, and you'll find hundreds of reptiles and amphibians, the quiet, watchful custodians of the forest floor and its streams. And that's before you even reach the water, where thousands of fish species paint the rivers and coral reefs with iridescent life.

But the true genius of this symphony lies in the almost invisible orchestra: the estimated 150,000 species of invertebrates. These are the tiny engines of existence—the insects, spiders, and worms that pollinate, decompose, and weave the very fabric of the ecosystem. So much of this world is still a mystery, a whispered secret. Scientists continue to find melodies we've never heard before, like the compound from Sarawak's Bintangor tree that once offered a glimmer of hope against the HIV virus. For centuries, indigenous communities have known how to read this living score, turning to nature's pharmacy for their well-being. This is not just a collection of species; it is a treasure chest of possibilities, a blueprint for survival.

A Tapestry in Tatters: The Slow Silence of a Forest

From a satellite's cold eye, Malaysia still looks green. The statistics boast that over half the country remains under tree cover, a figure that feels reassuring. But this number tells a comforting lie. It masks a deeper, more insidious wound: the slow, creeping death of fragmentation.

Imagine a magnificent, hand-woven tapestry, rich with a million colors and intricate designs. Now, imagine taking a pair of shears and cutting that masterpiece into a thousand small, disconnected squares. The threads are still there, but the picture is gone. The story is lost. This is what is happening to Malaysia's forests.

When a diverse, ancient rainforest is razed for a monoculture plantation like palm oil, it isn't replaced with another forest. It is replaced with a silent, green monotony. A natural rainforest is a multi-layered cathedral of life, with countless niches, food sources, and microclimates. A plantation is a biological desert. It has trees, but it has no soul.

For the wildlife caught within these shrinking islands of green, the world unravels. Species that need vast territories to hunt, mate, and live are suddenly prisoners in a cage with invisible bars. Their gene pool shrinks, a slow spiral into inbreeding and weakness that makes them vulnerable to disease and change. They are being silenced, one isolated population at a time.

The very air changes at the "edge" of these forest fragments. The deep, cool, moist interior gives way to a hotter, drier, and wind-blown boundary. It's a hostile environment for the specialists—the orchids, frogs, and insects that are tuned to the delicate frequency of the deep forest. They fade away, replaced by tougher, more common generalists.

This fragmentation is a death sentence delivered by a thousand cuts. It's the shy Malayan Tapir, a creature of the shadows, blinded by headlights on a new road that has sliced its home in two. It's the specialist kingfisher that can no longer find the specific stream-side habitat it needs to survive. It's the tragic echo of the Sumatran Rhinoceros, now gone from Malaysia's wild—a species whose last few individuals were scattered in fragments too far apart to ever find each other, their loneliness sealing their doom. Fragmentation doesn't just mean less forest. It means a sicker, weaker, and quieter world.

A Collision of Worlds: When the Wild Has Nowhere Left to Go

When you tear down the walls of a home, you cannot be surprised when the inhabitants wander out into the street, dazed and desperate. This is the heart of human-wildlife conflict in Malaysia. It is not an invasion by animals, but an expulsion. As their forest homes are shrunk and shattered, wild creatures are forced onto a collision course with humanity.

Elephants, with their long memories and ancient migratory routes, are the perfect, tragic example. They follow ghost-paths etched into their DNA, only to find the way blocked by a plantation. Driven by hunger, they enter these new landscapes, feeding on oil palm fruit. The result is a heartbreaking conflict: damaged crops, economic loss, and sometimes, tragic deaths on both sides. In states like Perak, the sheer volume of official complaints and compensation claims tells the story of this ongoing struggle.

The majestic Malayan Tiger, cornered in a shrinking pantry, is driven to hunt what it can find. When its natural prey disappears, it may turn to livestock, inviting the fear and fury of local farmers. A tiger that kills a goat is not a villain; it is a starving king in a collapsing kingdom.

Even the adaptable macaques, who seem to thrive on the fringes, are casualties. We see them as pests raiding garbage cans and snatching food, but their presence in our cities and towns is a sign of their own displacement. They are refugees, and our annoyance often leads to their persecution.

The conflict touches everything, from wild boars who raid farms to civets and snakes discovered in suburban backyards. Each encounter is a symptom of the same core disease: a world out of balance. For the animals, it leads to death by roadkill, poaching, or culling. For people, it brings fear and financial hardship, eroding the goodwill essential for conservation. This is the messy, painful friction that occurs when two worlds that should be separate are violently pushed together.

Weaving the World Anew: The Fight for a Connected Future

Amidst the scars of fragmentation, a new story is being written—one of hope, ingenuity, and a stubborn refusal to let the symphony fall silent. Malaysia is beginning to pioneer ways to stitch its tattered tapestry back together, balancing the needs of its people with the survival of its irreplaceable wildlife.

The most visible signs of this hope are the ecological corridors. These are not just bridges of concrete and steel, but bridges of life. Structures like the eco-viaduct at Sungai Yu are verdant lifelines arching over deadly highways, allowing elephants, tigers, and even smaller creatures to cross safely. They are a profound acknowledgment that the needs of wildlife are a part of our national infrastructure.

But the work goes deeper, into the soil itself. Across the country, especially in Borneo, conservation groups are undertaking the painstaking work of reforestation. They aren't just planting trees; they are resurrecting ecosystems. By planting a diverse mix of native species on land once scarred by plantations, they are rebuilding the pantries and nurseries for orangutans, pygmy elephants, and countless other species.

Crucially, this is not a battle fought by scientists alone. It is being won in partnership with the people who know the land best. Local and indigenous communities are being empowered as the frontline guardians of their natural heritage. They are becoming forest patrols, data collectors, and the keepers of community tree nurseries. Their ancient knowledge is not a relic of the past; it is the key to a sustainable future, providing livelihoods that protect, rather than destroy, the forest.

The work is multifaceted. It involves strengthening protected areas, using technology like camera traps and GPS collars to listen to the forest's needs, and creating compensation schemes that foster tolerance when conflicts do arise. It is a slow, complex, and delicate dance—a negotiation between economic reality and ecological necessity.

The future of Malaysia's wild heart is not guaranteed. It depends on the courage to see a healthy environment not as a luxury, but as the very foundation of a healthy nation. It requires us to move beyond simply counting trees and instead focus on the life that thrives between them. The goal is to transform the discordant noise of conflict and loss back into the rich, vibrant, and resilient symphony that is Malaysia's greatest treasure. The instruments are being mended. The score is being rewritten. We must all learn to listen.

The Two Hearts of Malaysia

To understand the story of modern Malaysia is to listen for two distinct heartbeats. The first is a powerful, metallic thrum—the relentless engine of an economy that has lifted millions, built gleaming cities, and carved a name for itself on the world stage. The second is an older, quieter pulse—the deep, green heartbeat of the rainforest, an ancient lifeblood that has nourished the land for millennia. For decades, the first heart has grown louder, stronger, often drowning out the second. This is the story of that dissonance, a tale of prosperity and loss written across the nation's changing landscapes.

The economic engine is fueled by two powerful currents: a golden river of palm oil and a torrent of timber. But drawing this power has scarred the land, leaving behind a complex legacy of environmental wounds, hidden financial risks, and a profound human cost.

The Golden Tide: Palm Oil's Complicated Reign

Palm oil is more than a crop in Malaysia; it's a destiny. As the world's second-largest producer, the nation's fate is inextricably tied to this versatile oil. It's the silent ingredient hiding in your morning snack, whispering from your shampoo bottle, and powering the biofuels in distant countries. This insatiable global appetite created a gold rush, transforming the very face of the nation.

Nowhere is this more true than on the island of Borneo. In Sabah and Sarawak, a journey that once would have passed under the dense, chaotic symphony of a rainforest canopy now reveals a different, more somber scene: endless, uniform rows of oil palm trees marching to the horizon. It is an image of immense efficiency and profound loss.

For every ton of golden oil, the forest pays a price. As primary forests are cleared, a world of irreplaceable biodiversity vanishes. The soulful eyes of the orangutan, the shy footfalls of the pygmy elephant, and the vibrant flash of exotic birds are replaced by a silent, green monoculture. The rich, tangled web of life, woven over eons, is unraveled thread by thread.

The cost is buried even deeper, in the very soil itself. Much of this expansion has occurred on peatlands, ancient, waterlogged soils that are not just dirt, but carbon libraries holding centuries of organic history. When these lands are drained for cultivation, it is like tearing out the pages of that library and setting them on fire. The released carbon billows into the atmosphere, making a commodity that seems so green a hidden driver of a warming planet.

There are whispers of a better way. Initiatives like the Roundtable on Sustainable Palm Oil (RSPO) offer a blueprint for a more conscious path forward. Yet, for many critics on the ground, these certifications are a quiet murmur against the roar of bulldozers. The central challenge remains a tightrope walk over a chasm: how can Malaysia nourish its economy without devouring its natural soul?

Echoes of the Chainsaw: A Legacy Carved from Timber

Long before the golden tide of palm oil, the sound that defined Malaysia's relationship with its forests was the roar of the chainsaw. The timber industry, both legal and illicit, was the original architect of change. Even "sustainable" selective logging, which cherry-picks the most valuable trees, can leave a wounded, fragmented ghost of a forest, disrupting the delicate balance of the ecosystem.

More brutal is the practice of clear-felling, which erases the forest entirely to make way for industrial timber plantations. These neat rows of fast-growing trees are farms, not forests. They lack the spirit, the complexity, and the resilience of the ecosystems they replaced.

Disturbingly, investigations have uncovered a cynical shell game. State-sanctioned projects, ostensibly for agriculture, are sometimes used as a cover for a one-time timber grab. Once the ancient, valuable trees are felled and sold, the promise of a thriving plantation fades, leaving behind a barren landscape and questions of governance and corruption.

This industry built livelihoods. It put food on the table in rural communities and its products built homes across the world. But it was a boom financed by a finite inheritance. Now, with natural forests depleted, the nation confronts the hard truth of a resource consumed. The shift to plantations is an attempt to create a sustainable future, but these sterile monocultures can never replicate the life-giving services of a true forest. The challenge is to transform an industry of extraction into one of stewardship.

The Bill Comes Due: Nature's Unseen Balance Sheet

For a long time, the degradation of nature was seen as an unfortunate but separate side effect of economic growth. That illusion is shattering. The environmental cost is now showing up on the nation's financial ledger as a staggering risk.

Nature, it turns out, keeps its own books. And the bill is coming due. A landmark World Bank study sent a shockwave through the country, projecting that a "partial collapse of ecosystem services" could erase 6% of Malaysia's GDP annually by 2030. This isn't an abstract fear; it's the tangible cost of losing the forest's gift of clean water, the soil's ability to nurture crops, and the local climate's stability. When the land is sick, the economy catches the fever.

Malaysia's own central bank, Bank Negara Malaysia, has sounded the alarm. It found that an astonishing 54% of all commercial loans are exposed to the physical risks of nature loss. This means the stability of the entire financial system is tethered to the health of its ecosystems. The farmer whose crops fail due to drought, the tourist resort left stranded by a degraded coastline, the fishery with no fish—their financial troubles ripple upward, shaking the very foundations of the banks that fund them.

Malaysia's economy was built on a bedrock of immense natural capital. For decades, it has been drilling into that foundation for short-term gain. Now, the cracks are beginning to show.

Guardians of the Green: The Human Price of Progress

Beyond the balance sheets and biodiversity counts lies the most profound cost of deforestation: its impact on people. For Malaysia's Indigenous communities, the forest is not a resource to be exploited. It is an ancestor, a pharmacy, a library, and a cathedral. Their cultures, identities, and very survival are woven into the land they have inhabited for generations.

The story of the Iban community in Sarawak is the story of so many. They stand as guardians at the edge of the disappearing forest, facing down logging companies with little more than their ancestral claims. In return, they receive threats and eviction orders. Their struggle is a microcosm of a global conflict, where the ancestral lands that are the heart of their identity are labeled as commodities on an international supply chain. The wood from their home may end up as a piece of furniture in a distant living room, its origins and true cost unknown.

When the bulldozers arrive, they lose everything. Their hunting grounds, their sources of traditional medicine, their sacred sites, and the graves of their ancestors are all erased. It is a cultural unraveling, a forced displacement that severs the bond between generations and silences ancestral stories forever.

Their fight for recognition is a fight for the soul of the country. To protect their rights is not just a matter of justice; it is a recognition that their traditional wisdom holds the keys to the sustainable future Malaysia seeks. They are not standing in the way of progress; they are pointing the way toward a more balanced path.

In the end, Malaysia stands at a crossroads. The thrumming heart of the economy and the ancient pulse of the forest need not be at war. The challenge is to learn to listen to both heartbeats at once, to forge a new rhythm of progress—one that is not only prosperous but also just, resilient, and whole. The choice is about more than protecting forests; it's about deciding what kind of ancestor Malaysia chooses to be.

In the heart of Southeast Asia, Malaysia is learning a new dance. It's a delicate, high-stakes tango between the thrumming pulse of its economy and the quiet, ancient rhythm of its rainforests. Blessed with a staggering wealth of biodiversity yet anchored to the global demand for commodities like palm oil, the nation finds itself on a tightrope, striving for a balance that feels, at times, almost impossible. This is a story of grand promises etched into policy, of global handshakes, and of the stubborn shadows that persist between intention and reality.

The National Handshake: A Promise Called MSPO

At the center of this story is a promise, a national handshake known as the Malaysian Sustainable Palm Oil (MSPO) certification. Born from a global chorus of concern—whispers in consumer aisles and shouts from environmental watchdogs that Malaysia could no longer ignore—MSPO is the nation's answer. It's a pledge that the golden oil flowing from its shores will not cost the Earth its lungs or its people their dignity.

What began as a voluntary suggestion in 2013 became a national mandate by the end of 2019. It was a line drawn in the sand, a signal that sustainability was no longer a choice but the very fabric of the industry. This promise is woven from three core threads: protecting the environment, honoring social responsibility, and ensuring economic resilience.

The promise grew stronger, evolving. As of January 1, 2025, the updated MSPO 2.0 standard carries a powerful new clause: a strict deforestation cut-off date of December 31, 2019. This is more than a rule; it's a declaration. Any forest cleared after this date cannot produce certified palm oil. It's a lock on the door to future deforestation, an attempt to tell the world that growth will no longer come at the expense of its irreplaceable wild spaces.

This updated vow goes deeper, standing guard over High Conservation Value (HCV) areas—the sacred groves, the peatland hearts, the last refuges of endangered life. It outlaws the choking haze of open burning, a ghost of land-clearing practices past. And it extends a hand to the workers in the fields, demanding fair wages, safe conditions, and an end to exploitation.

The response has been remarkable. By April 2024, over 87% of Malaysia's palm oil landscapes were waving the MSPO flag. It's a testament to a monumental effort to turn the ship around. Yet, the journey isn't over. That remaining percentage, often composed of small, independent farmers, represents the final, most challenging miles. And while MSPO is Malaysia's own standard, tailored to its unique soul, critics wonder if its heartbeat is as strong as other global certifications like the RSPO. The debate continues, but MSPO remains the foundational stone upon which Malaysia is building its sustainable future.

A Voice on the World Stage

Malaysia's quest for balance doesn't stop at its coastline. It has stepped onto the global stage, its voice joining the international choir singing for the planet's future. As a signatory to the Convention on Biological Diversity (CBD), Malaysia is part of a worldwide pact to protect the very web of life.

Now, the nation is weaving the threads of the new Kunming-Montreal Global Biodiversity Framework into its own national policy. It's a commitment to a breathtakingly ambitious goal: to halt and reverse the loss of nature by 2030. Central to this is the "30x30 target"—a global pledge to protect 30% of the world's land and sea within the decade. Malaysia has raised its hand in support, and crucially, its people are standing right behind it. An astounding 96% of Malaysians back this goal, their collective voice a powerful mandate for bold action. This isn't just policy; it's a popular movement, a shared dream for a greener nation.

Beyond national parks and declared reserves, Malaysia is also looking for conservation in unexpected places. It is pioneering a framework to recognize "Other Effective Area-Based Conservation Measures" (OECMs). These are the unsung heroes of conservation: ancestral lands managed by indigenous communities, sacred natural sites, and even farmland that nurtures biodiversity. It's a beautiful, holistic recognition that protecting nature isn't just about fencing it off; it's about acknowledging the stewardship that already exists in the heart of communities.

The Ghost in the Machine: Where Policy Meets a Wall

For all the beautifully crafted policies and international pledges, a ghost haunts Malaysia's sustainability efforts: the gap between paper and practice. It's a chasm where weak enforcement and faltering political will can allow the spirit of the law to wither.

Even under the watchful eye of MSPO, shadows can linger in the plantations. Reports from the ground reveal that a certificate on the wall doesn't always guarantee that every root and every stream is protected. The system sometimes suffers from a lack of vigilant oversight, with too few eyes to watch over vast territories. Penalties can feel less like a deterrent and more like a slap on the wrist, allowing "bad actors" to treat fines as just another cost of doing business.

This challenge is complicated by a kind of bureaucratic tangle. Imagine different government ministries as skilled musicians in an orchestra, each with their own sheet music. When the Ministry of Agriculture, the Ministry of Environment, and the bodies governing land use don't play in perfect harmony, the result is dissonance. One hand may approve a permit while the other decries its environmental impact. This fragmentation can slow action to a crawl, allowing damage to occur in the silence between agencies.

Ultimately, the ghost is sustained by a wavering of political will. The strongest laws are merely ink on paper without the unwavering courage to enforce them, especially when powerful economic interests are at play. Environmental watchdogs continue to point to this chasm, where the reality on the ground—stubborn deforestation, lingering labor issues—doesn't quite match the soaring rhetoric of sustainability pledges. Closing this gap is not just a matter of logistics; it's a matter of heart and political courage.

The Paradoxical Pulse of Money

Money is the lifeblood of any national endeavor, and in Malaysia, its pulse reveals a deep paradox. On one hand, the nation is taking innovative steps to fund its green ambitions. Through the UNDP's BIOFIN initiative, Malaysia created a national Biodiversity Finance Plan, a roadmap to heal a significant funding gap for conservation. It's a creative treasure hunt for new resources, exploring everything from green bonds and ecotourism levies to payments for the clean air and water that healthy ecosystems provide.

Yet, as this green-tinted money trickles in, a torrent of capital flows in the other direction. Data from watchdogs like Forests & Finance tells a startling story. Malaysian financial institutions are world leaders in funding the very industries that put forests at risk. The nation ranks second globally in holding stocks and bonds in these sectors, to the tune of a staggering $6.9 billion. It also sits seventh in providing credit.

This is the central paradox: a nation striving for sustainability with one hand, while its other hand signs checks that could finance deforestation. It's as if the country has a split personality. The nation's financial heart is not yet fully aligned with its environmental conscience. This flow of "risk capital" into palm oil, timber, and paper production threatens to undermine the hard-won progress of MSPO and other green initiatives. Greening the financial sector itself—steering investment away from destructive practices and towards regenerative ones—is perhaps the most profound challenge of all.

An Unfolding Story

Malaysia's sustainability story is not one of simple triumph or failure. It is a complex, evolving narrative of a nation grappling with its identity in the 21st century. The MSPO certification is a powerful statement of intent, and its embrace of global conservation goals, backed by the resounding support of its people, shows a genuine desire for change.

Yet, the journey is fraught with real and persistent obstacles. The path forward demands more than just well-meaning policies; it requires the muscle of enforcement, the harmony of governance, and a deep, systemic shift in the way money flows. Success will be measured not just by the percentage of certified land, but by the health of the forests, the well-being of the workers, and the courage to make the hard choices. Malaysia's dance continues, a compelling, vital, and unfinished performance on the world stage.

Philippines - Marine Degradation & Land Use Change

The Sea's Silent Scream

Imagine an underwater kingdom painted in a thousand impossible colors. This is the Coral Triangle, the planet's cradle of marine life, and the Philippines is its vibrant heart. But the symphony of life here is being drowned out by a deafening silence. More than 75% of the nation's fishing grounds are gasping for air, their lifeblood drained by a relentless human thirst.

For years, a brutal war has been waged in these waters. The thunder of dynamite fishing shatters coral castles that took centuries to build, leaving behind nothing but rubble and the ghosts of the fish it killed. It's not fishing; it's annihilation. Then there's the whispered poison of cyanide, a chemical fog that drifts through the reefs, stunning fish for the aquarium trade but leaving a trail of death in its wake. It chokes the coral polyps, the tiny architects of this underwater world, until the reef itself turns into a bleached skeleton.

These aren't just statistics; they are wounds. They are the empty nets of artisanal fishers whose calloused hands now pull up less and less. They are the loss of a natural sea wall that once protected coastal villages from the rage of the sea. When a reef dies, a part of the nation's soul dies with it.

Now, a global fever is setting in. Climate change warms the ocean, forcing the corals to expel their life-giving algae in a desperate, ghostly white surrender known as bleaching. The water itself is becoming more acidic, a slow-motion dissolving of the very shells and skeletons that form the foundation of this marine world. The silent scream of the sea is growing louder.

A Scar on the Soul of the Land

The crisis doesn't end at the shoreline. On land, a great green cloak is being torn to shreds. Since 2000, the Philippines has lost a staggering 1.52 million hectares of tree cover—an 8.2% decrease that has left a scar on the nation's soul. This isn't just about trees; it's about the erasure of homes for countless creatures that are the Philippines' unique living heritage.

Today, at least 700 species are threatened, their existence hanging by a thread.

The majestic Philippine Eagle (Pithecophaga jefferyi), king of the skies, now finds its kingdom shrinking, with fewer than 800 individuals left to soar over forests that are disappearing.

The fierce yet vulnerable Tamaraw (Bubalus mindorensis), a wild buffalo found only on Mindoro, is being cornered by expanding farms and the poacher's gun.

The brilliant Philippine Cockatoo (Cacatua haematuropygia) is being silenced, its forest homes bulldozed and its beauty coveted for the illegal pet trade.

This destruction is driven by the relentless hunger of illegal logging, the gaping wounds of mining operations, and the steady creep of farms and cities that devour the wild. Each fallen tree, each cleared acre, pushes these irreplaceable species closer to becoming mere legends whispered to future generations.

A Fevered Sky, An Angry Sea

The Philippines has always lived with typhoons, a nation forged by wind and water. It lies in the heart of the Pacific's typhoon belt, weathering an average of 20 storms a year. But climate change has turned this familiar struggle into a terrifying new reality. It's a threat multiplier, pouring fuel on an already raging fire.

The storms are becoming monsters. Super typhoons, once rare, now lash the islands with unimaginable fury, their winds stripping forests bare and their waves smashing what's left of the coral reefs. They are nature's grief, made more powerful by a warming world.

At the same time, the ocean is slowly, relentlessly rising. A quiet thief, it steals inches of coastline, submerging low-lying islands, poisoning freshwater with salt, and drowning the vital mangrove forests that serve as the sea's nurseries. The weather itself has become a wild, unpredictable beast, swinging from scorching droughts that crack the earth to torrential floods that wash away lives and livelihoods.

A Chorus of Hope

This is not just an ecological tragedy; it is a human one. It is the story of coastal communities whose pantries are empty, of farmers whose lands are broken, and of indigenous cultures whose identities are woven from the very nature that is vanishing.

Yet, in the face of this immense crisis, the Filipino spirit endures. This is not a story of defeat. It is a call to arms.

Across the archipelago, a chorus of hope is rising. Communities are banding together to guard their marine sanctuaries. Reforestation programs are planting seeds of a new future. Brave individuals are standing up to illegal loggers and miners. The Philippines is a key partner in the Global Biodiversity Framework, a worldwide promise to protect 30% of our planet's land and sea by 2030.

The fight to save the Philippines' natural wonders is a fight for its very identity and for the future of its people. It is about honoring a sacred trust and ensuring that the archipelago's heartbeat—once faint and fading—grows strong and steady once more for all the world to hear.

The story of the Philippines' vanishing beauty is not written in the language of ecology alone. It is written in the lines on a fisherman's face, in the gnawing hunger of a child, and in the choices a family must make when survival is the only law. The wounds on this archipelago of 7,000 islands are not just environmental; they are deeply, tragically human. They are born from the tangled realities of poverty, a burgeoning population, and a society struggling to find its footing.

A Requiem for the Reefs

Listen. Can you hear it? A muffled boom beneath the waves. It's the sound of desperation. For a fisherman with an empty net and an emptier stomach, it's the sound of a quick, bountiful catch. For the ocean, it is a heartbeat stopping.

This is dynamite fishing. A single blast shatters a city of coral built over centuries, turning it into a bone-white graveyard. It's like setting a forest ablaze to hunt a single deer. In other places, a cloud of cyanide drifts through the water, a silent, ghostly assassin. It stuns the brilliantly colored fish destined for foreign aquariums, but it suffocates the life from the corals they call home.

These are not the acts of villains, but of fathers pushed to the brink. They are caught in a cruel paradox: to feed their family today, they must destroy the very source of tomorrow's food. As the reefs die, the fish vanish. To catch what little remains, they must resort to ever more desperate measures, deepening the wound. It is a spiral of destruction that drains the sea of its life and the people of their future.

The cost is not just ecological. It's a staggering economic hemorrhage, an estimated loss of $1.2 billion over two decades that could have built schools and healed the sick. But the truest cost is measured in empty plates and hollowed-out futures. For millions living along the coast, the sea is their grocery store, their employer, and their identity. As the fish disappear, a protein vital to a child's growth is replaced by hunger. The songs of the sea, the culture woven from generations of fishermen, begin to fade into a sorrowful silence.

The Weeping Mountains

The lament of the sea finds its echo in the highlands. The forests, the great green lungs of the islands, are gasping for breath. Once, misguided policies handed out logging permits like short-term leases, inviting companies to clear-cut ancient trees with no thought for tomorrow, leaving behind a scarred, bleeding earth.

Today, the wounds are inflicted by a thousand smaller cuts. A family, with no land to their name, carves a small patch from the forest to plant enough to survive. Elsewhere, vast corporate plantations march up the hillsides, bulldozing forests for seas of palm oil and pineapple, shipping the land's fertility overseas. In the mountains, the glint of gold and copper invites mining companies to rip the earth open, creating moonscapes where life once thrived, poisoning rivers with a toxic legacy that will outlive the profits.

Cities swell, spilling concrete and asphalt over what was once green. Roads slice through ancient wilderness, and dams drown valleys, silencing the life within them. And in the shadows, illegal loggers continue their work, their chainsaws the percussion in a symphony of destruction.

This loss is felt most acutely when the skies open. Without the tangle of roots to hold it, the soil melts away in the rain. Rivers, no longer gentle, become raging monsters of mud and debris. They sweep through villages, carrying away homes, hopes, and lives. The forest's death becomes the valley's flood.

A Poison in the Veins

A silent killer seeps through the nation's veins. On land and at sea, pollution is the slow, creeping sickness that underlies it all. From farms, a chemical rain of pesticides and fertilizers washes into the rivers, creating suffocating algal blooms that choke the life from the water. From factories and cities, untreated sewage and industrial waste pour directly into the sea, a toxic cocktail for marine life.

But the most haunting plague is plastic. The Philippines, an island nation, finds itself drowning in a tide of its own making, becoming the world's largest source of plastic in our oceans. It is a crisis of convenience and circumstance. A landscape lacking in waste disposal infrastructure, a culture of single-use sachets, and a geography of countless rivers and coastlines all conspire to turn the surrounding seas into a plastic soup. It's a tragedy you can see on every shore, but the greater danger is the invisible tide of microplastics entering the food chain, from the smallest fish to the people who eat them.

The Human Price of a Dying World

The collapse of these natural worlds is, in the end, the collapse of human worlds. It is a story about vulnerability.

When the coral reefs are gone, the ocean's rage comes ashore unchecked. When the mangrove forests—the coast's gnarled, steadfast guardians—are cut down, a typhoon's storm surge doesn't knock; it walks straight into the living room. The Philippines is a nation born in the path of storms, but this man-made devastation has stripped away its natural armor. Disasters that were once survivable become catastrophic, costing the economy an average of $0.5% of its GDP every year, a tax paid in ruined infrastructure and shattered lives.

And this price is not paid equally. It is the poor who are hit hardest. They are the ones living in flimsy homes on the vulnerable riverbanks and low-lying coasts. They are the ones with no savings, no insurance, and no safety net. A single typhoon, a single flood, can erase a lifetime of work, plunging a family into a depth of poverty from which they may never escape.

This is the ultimate consequence. A dying environment doesn't just mean lost species and barren landscapes. It means hunger. It means sickness from poisoned water. It means communities displaced and cultures unraveled. It means a mother in an evacuation center, wondering how she will rebuild a life from the mud and debris of what was once her home.

To heal the Philippines, we must look beyond the symptoms—the dead reef, the barren mountain, the plastic-choked river. We must look into the heart of the matter: the human struggle for dignity, security, and a place to call home. The health of the Filipino people and the health of their extraordinary islands are not two separate issues. They are one and the same story, waiting for a new, hopeful chapter to be written.

The Archipelago's Green Heartbeat: A Symphony of Survival

In the Philippines, the earth speaks in a language of unparalleled life. It's in the whisper of the emerald canopy, the flash of a tarsier's eye, the silent dance of a whale shark in sapphire depths. This archipelago of over 7,000 islands is not just a country; it is a living, breathing symphony of biodiversity. But today, that symphony is playing in a minor key, threatened by a chorus of challenges. Yet, rather than let the music fade, the Filipino people and their government are composing a new score—one of bold promises, fierce resolve, and ingenious creativity.

A Vow Echoed by Millions

This isn't just policy; it's a promise made on the world stage. By joining the High Ambition Coalition for Nature and People, the Philippines has vowed to protect 30% of its land and seas by 2030—the globally renowned "30x30 target." This is more than a statistic; it's a sacred pact with the future, a declaration that the nation's natural soul is not for sale.

What transforms this promise from a political gesture into a national crusade is the voice of the people. Imagine a cause so profound that it unites 95% of a nation. A 2022 poll revealed this staggering consensus of support for the 30x30 goal. This is not a top-down mandate; it is a grassroots groundswell, a powerful current of public will that gives leaders the courage to navigate the turbulent waters of change. It is the nation's heart beating in unison for its natural heritage.

This collective will finds its blueprint in the Kunming-Montreal Global Biodiversity Framework (GBF). The Department of Environment and Natural Resources (DENR) is tasked with translating this global vision into local reality. Through nationwide workshops, they are weaving the 23 targets of the GBF into the very fabric of the Philippine Biodiversity Strategy and Action Plan (PBSAP). This is a "whole-of-society" approach, a recognition that the forest's health is linked to the farmer's choice, the reef's survival to the tourist's footprint, and the child's future to the lessons they learn today.

The Great Financial Chasm

But what is a promise without the means to keep it? Herein lies the elephant in the room, the silent chasm that threatens to swallow these grand ambitions. To truly breathe life into this new plan, the Philippines needs around PHP 24 billion ($400 million) every year. This is the real cost of guardianship—of funding the rangers who walk the lonely patrols, the scientists who monitor fragile ecosystems, and the projects that restore life to scarred lands and seas.

The current reality? Only PHP 5 billion ($85 million) finds its way to the front lines of conservation. This is not a funding gap; it's an 80% canyon of silence. It's the sound of a protected area without enough guardians, of environmental laws without teeth, of restoration plans that remain beautiful dreams on paper. This chasm threatens to turn the nation's symphony of survival into a requiem.

Composing a New Economy for Nature

Faced with this daunting silence, the Philippines, with partners like the UNDP's BIOFIN, is refusing to despair. Instead, it is getting creative. It is composing a new economic melody where nature's value is finally recognized on the balance sheet. This is a story of ingenuity, a mosaic of solutions designed to bridge the chasm.

The first notes were played by the people themselves. When the pandemic silenced tourism and left the brave rangers protecting the critically endangered tamaraw without pay, a crowdfunding campaign, "Together for Tamaraws," rose to the occasion. It was a powerful demonstration of solidarity, proving that a thousand small contributions can create a lifeline of hope.

Then, innovation went digital. A mobile game, "Animal Town," was born, transforming screen time into a force for good. Players learn, engage, and through in-app purchases, contribute directly to conservation funds. It's a brilliant fusion of modern entertainment and ancient responsibility, inviting a new generation to become digital guardians of biodiversity.

The orchestra of solutions is growing more complex and powerful. The nation is looking at the very systems that have caused harm and asking, "How can we make them heal?" Imagine redirecting agricultural subsidies that encourage chemical-heavy farming and instead, rewarding farmers who cultivate resilient heirloom rice varieties or embrace organic methods. This is not just about stopping harm; it's about actively funding a partnership between those who feed the nation and the land that sustains them.

This new composition includes a suite of other innovative instruments waiting to be played:

Ecotourism Levies: A small fee from every visitor, creating a direct flow of revenue from those who enjoy nature back to those who protect it.

Green Bonds: Tapping into capital markets to fund large-scale environmental projects, allowing investors to put their money where their values are.

Payment for Ecosystem Services (PES): A simple, powerful idea where communities upstream who protect a watershed are paid by the cities downstream that receive clean water.

Debt-for-Nature Swaps: A creative financial arrangement where a portion of national debt is forgiven in exchange for commitments to invest in conservation.

The Unfinished Symphony

The Philippines stands at a crossroads, armed with an unbreakable public mandate and a portfolio of brilliant ideas. The challenges of political shifts, climate change, and enforcement are real and formidable. But the nation's resolve is a powerful counter-melody.

The journey ahead is about more than just finding money. It's about rewriting our relationship with the natural world. It is about proving that a thriving economy and a thriving ecosystem are not opposing forces, but two parts of the same beautiful harmony. The archipelago's symphony of survival is still being written, and with every creative solution and every act of collective will, the music grows stronger, more hopeful, and more profound. The world is listening.

China - Infrastructure, Consumption & Policy Innovation

China is a nation with a split soul. One soul is that of the poet and the painter, the one who saw sublime power in the mist-shrouded peaks of Huangshan and wisdom in the quiet rustle of a bamboo grove. This is the soul of a civilization that lived in conversation with its landscape for millennia. The other soul was born in the fever dream of the 20th century, a soul of steel and concrete, convinced that progress is measured in smokestacks and the defiant glow of cities that never sleep. For decades, this second soul has been winning.

To stand in modern China is to feel this conflict under your feet. A high-speed train, a silver dragon of unimaginable speed, slices through a valley where, generations ago, tigers roamed. Beneath the foundations of a shimmering skyscraper in Shenzhen lies the ghost of a wetland, its memory paved over by ambition. This isn't just development; it's a form of geological amnesia. The nation's great rivers, the very arteries that nourished its ancient heart, now often run thick with the runoff of a million factories, their waters a toxic brew that has left a third of the rural populace thirsting for a clean drink. In Shanghai, the ground itself gives way, sinking millimeter by millimeter under the weight of its own success, a slow-motion collapse that is both literal and spiritual.

This great silencing has echoed loudest in the wild. Imagine the forests and grasslands of a nation that holds 14% of the world's vertebrate life, now quietly emptying. Nearly 70% of its mammals are ghosts in their own land, their numbers dwindling, their futures written in red ink. The last dance of the stunning Green Peacock (Pavo muticus) plays out in shrinking patches of forest, its stage threatened by the hum of a future hydroelectric dam. The conflict is stark: the energy to power a city versus the last space on Earth for a bird of mythic beauty.

Then there is the hunger. A voracious, global appetite for the rare and the exotic finds its epicenter here. This hunger fuels a shadow economy of immense cruelty, turning living beings into commodities. The pangolin, a shy, scaled creature of the night, has become the world's most heartbreaking product. Hunted for its scales, believed to hold medicinal power, and for its meat, a fleeting status symbol on a dinner plate, it is being erased from the planet. This demand reaches far beyond China's borders, a destructive echo that empties forests across Asia and Africa.

And yet, just as the abyss seems absolute, the poet's soul begins to stir. A political seismic shift is underway. The same monolithic state that drove the bulldozers is now attempting a course correction of breathtaking scale. The Ecological Conservation Redline is not just a policy; it is an act of national will, a line drawn across the map declaring vast territories sacred and untouchable. It is a Great Wall of conservation, built not of stone, but of political resolve, designed to hold back the tide of destruction.

In secretive breeding facilities, there are glimmers of this new hope. The primeval Sichuan taimen (Hucho bleekeri), a giant salmon fighting for its life, is being raised with the dream of returning it to its ancestral rivers. In the cloud-wrapped mountains of the interior, conservationists watch over the ethereal golden snub-nosed monkey (Rhinopithecus roxellana). These are acts of ecological penance, attempts to stitch back together what has been so violently torn.

But the scars run deep. A 2025 biodiversity report from Hong Kong serves as a warning etched in miniature: even in a city famed for its country parks, the monster of development is never fully tamed, with over a quarter of its species facing extinction.

Herein lies the epic drama of our time. China's top-down power is both a blessing and a curse. It is the only force capable of enacting change on such a colossal scale, yet it can also be a blunt instrument, deaf to the whispers of local communities whose cooperation is the bedrock of true, lasting conservation. But something new is happening. Green shoots are pushing through the concrete. A vibrant community of local NGOs, citizen scientists, and a newly aware public are adding their voices to the fight.

The world watches, not just as a spectator, but as a stakeholder. China's story is our story. The fate of countless species, the health of the global climate, and the very possibility of a sustainable future are being decided there, now. The question that hangs in the polluted air and the protected forests is a simple one: Which of its two souls will shape the twenty-first century?

A Ghost in the Machine: The True Price of China's Miracle

For decades, the world has watched in awe as China performed an economic miracle, pulling a nation from hardship into the light of prosperity. It's a story told in skylines of impossible steel and light, in bustling ports, and in the quiet hum of factories that became the world's workshop. But beneath the roar of this engine, another story has been unfolding—a ghost ledger keeping track of a different kind of debt. This is the story of how the very forces that built modern China are systematically unmaking the natural world, a bill that is now coming due not just in remote forests, but in the heart of its economy and the health of its people.

The Dragon's Concrete Veins

The most visible symbol of this bargain is the infrastructure itself. China's ambition is written in concrete and steel. At home, cities have metastasized, their sprawl consuming ancient wetlands and forests, silencing ecosystems that had thrived for millennia. But this domestic transformation was only a prelude. With the launch of the Belt and Road Initiative (BRI), the dragon stretched its limbs across the globe.

More than a network of roads and rails, the BRI is a system of new concrete veins being laid over the planet's most vital, life-giving organs. These corridors are not drawn on empty maps; they slice through the Earth's richest tapestries of life—the breathing rainforests of Southeast Asia, the soaring mountains of Central Asia, the vibrant savannas of Africa.

The first cost is paid in the violent act of construction: forests fall, mountains are hollowed out, and wild rivers are tamed into silent reservoirs. But the deeper wound is the fragmentation that follows. A continuous, thriving wilderness becomes a collection of lonely islands of green, marooned in a sea of development. For countless animals, this is a slow-motion death sentence. The paths to find a mate, a meal, or fresh water are severed. Gene pools shrink, and the shadow of extinction lengthens. These new roads also become expressways for destruction, granting poachers, illegal loggers, and miners easy access to previously untouchable sanctuaries.

Consider the story of the Tapanuli orangutan in Sumatra, a quiet tragedy playing out in the shadow of a BRI-funded dam. Here, in the only home this family of great apes has ever known, a new power station threatens to shatter their world. With fewer than 800 individuals left, scientists warn that this single project—a small knot in the vast BRI network—could deliver the final, fatal blow, pushing our flame-haired cousins into the silence of forever. Their plight is a postcard from the front lines, one of thousands of species caught in the path of a development model that sees the Earth as a resource to be conquered, not a home to be shared.

The World's Appetite, The World's Ghost Forests

China's role as the world's primary manufacturer and consumer has created a voracious, global appetite. The nation's factories and markets demand a constant flow of raw materials, and in doing so, China has effectively begun to "import" extinction, outsourcing the environmental cost of its consumption to distant lands.

Follow the trail of a piece of luxury rosewood furniture, prized in a Shanghai high-rise. Its journey likely began with the illegal felling of a tree in Madagascar's unique spiny forests, a theft that robs the island of its irreplaceable biodiversity and pushes its iconic lemurs closer to the edge. Follow the demand for animal feed that fattens the pork and poultry to feed a nation. That trail leads directly to the ashes of the Amazon and the plowed-under savannas of the Cerrado, where vast, biodiverse ecosystems have been sacrificed for seas of soy monoculture. Follow the palm oil, a hidden ingredient in countless products, and you arrive in the decimated rainforests of Indonesia and Malaysia, where the habitats of orangutans, tigers, and elephants have been cleared for plantations.

This is the ghost in the global machine. Through the tangled, opaque webs of international supply chains, the environmental violence is sanitized. The consumer in China sees a finished product, the corporation sees a profit margin, and the world sees economic growth. But the ecological price is paid in full by source countries, often those least able to bear it, leaving behind a legacy of poisoned rivers and silent forests.

A Harvest of Consequences

Back within its own borders, the strain of feeding its people has turned the land against itself. To squeeze astonishing yields from just 7% of the world's arable land, China unleashed a chemical blitzkrieg on its farms. For years, an avalanche of fertilizers and pesticides saturated the soil. The runoff has bled into rivers, creating monstrous algal blooms that choke the water, starve it of oxygen, and create vast "dead zones" where nothing can live. The pesticides, while killing their targets, have waged a quiet war on the allies of agriculture—the bees needed for pollination and the vital microorganisms that give soil its life.

This agricultural assault was amplified by an industrial one. The skies of the industrial heartlands filled with smog so thick it became a global headline, while heavy metals and toxins from unchecked factories bled into the waterways. Along these poisoned rivers, the whispers started, then grew into a national outcry about "cancer villages," communities where sickness and death were a horrifyingly common price for living near a factory.

The bill has arrived. It is not an abstract ecological warning; it is a clear and present danger to China's own prosperity. The World Economic Forum delivered a shocking diagnosis: 65% of China's GDP is threatened by the collapse of the natural systems that provide it with clean water, fertile soil, and breathable air. The polluted water affects millions, the smog is linked to countless premature deaths, and the degraded land threatens the nation's food security.

In a remarkable turn, the government that presided over this destruction is now launching some of history's most ambitious healing projects. The "Grain for Green" initiative, a grand act of ecological penance, has paid farmers billions to let their fields return to forest and grassland, measurably increasing the country's green cover.

Yet, even here, a painful human truth emerged. When the subsidies ran out, many farmers, left with no other way to survive, had to cut down the new saplings and plow the land once more. It was a heartbreaking lesson: you cannot heal the land without healing the lives of the people who live on it. True restoration isn't about planting trees; it's about planting new, sustainable livelihoods that allow communities to become guardians of their environment, not its casualties.

China stands at a profound crossroads. The dragon of its economy is waking up to the fact that it has been devouring its own tail. The path forward demands more than technological fixes or top-down decrees. It requires a fundamental reimagining of value, a new consciousness where a thriving forest is seen as more precious than a new highway, and clean water is understood as the ultimate currency. The story of China's environmental reckoning is not just its own; it is a lesson for a world built on the myth of infinite growth on a finite planet. The true measure of a nation's strength is not just in what it can build, but in what it has the wisdom to preserve.

A Tale of Two Chinas, and the Great Healing

For a generation, the world has watched China build. We saw the impossible skylines pierce the clouds, the bullet trains stitch together a continent, the factories hum with a power that reshaped the global economy. This was the China of steel and speed, a nation running a breathless marathon toward modernity. But behind this story, another was unfolding—a quieter, more desperate tale whispered in the smog-filled air and reflected in the murky waters of its great rivers. It was the story of a land gasping for breath.

Now, the marathon runner has stopped, looked back at the exhausted landscape, and chosen a radically new path. China is undertaking an environmental course-correction so massive, so audacious, it defies easy comparison. This isn't just about planting trees or starting recycling programs. It's a fundamental rewiring of the nation's soul, powered by the full, terrifying force of a centralized state. It is the story of a nation trying to heal itself before it's too late.

The genesis of this shift isn't a grassroots movement; it's an imperial decree for the 21st century. The concept of "Ecological Civilization" (she ¯ ngta ` i we ´ nmı ´ ng), championed by President Xi Jinping and etched into the constitution, is the philosophical engine. It recasts the natural world not as a resource to be plundered, but as the very foundation of national strength and destiny. It's a profound declaration: the mandate of heaven now includes clean air and living rivers.

Drawing a Line Around the Nation's Heart

The most potent symbol of this new mandate is the Ecological Conservation Redline (ECRL). Forget everything you know about traditional national parks. Imagine, instead, a doctor identifying the vital organs of a patient—the heart, the lungs, the arteries—and drawing an unbreachable line of protection around them. This is what China has done to its own map.

With scientific precision, the state has identified the lands and waters that are non-negotiable for its survival. We're talking about the towering headwaters of the Yangtze and Yellow Rivers, the "water towers" for half a billion people. We're talking about the misty mountain sanctuaries of the last giant pandas and the elusive snow leopards. We're talking about the coastal mudflats that serve as the Grand Central Station for millions of migratory birds.

The scale is almost incomprehensible: a territory larger than India, where the bulldozer must fall silent. Today, at least a quarter of China is inside this protective embrace. The redline is not a suggestion; it is an ultimatum. When a major dredging project threatened the delicate ecosystem of Shenzhen Bay—a crucial fueling stop on a bird migration flyway stretching from Australia to the Arctic—the redline brought it to a dead halt. The message was clear: within these sacred zones, the economy no longer gets the deciding vote. This single, bold policy is China's primary tool for meeting its promise to the world: to help protect 30% of the planet by 2030.

Of course, this is no simple task. This redline cuts across lives and livelihoods. It means moving factories, rerouting railroads, and asking communities to fundamentally change their relationship with the lands they have always known. The response has been a massive, complex effort to create what are called "ecological compensation" programs—finding ways to pay people to be guardians of the land, turning the act of protection into a new kind of economy.

From the Back of the Room to the Head of the Table

For decades, at global environmental summits, China often sat in the back, a reluctant giant wary of commitments that might slow its economic engine. That China is gone. The turning point was its presidency of the UN's COP15 biodiversity summit. Stepping into the spotlight, China became the indispensable mediator, the nation that brokered the deal when talks faltered. The result was the Kunming-Montreal Global Biodiversity Framework (GBF)—the world's new master plan for pulling nature back from the brink.

Instead of taking a victory lap, China rushed home and began the hard work. It was one of the first nations to draft a detailed national strategy spelling out exactly how it would meet the 23 ambitious targets of the GBF. This isn't just a job for the environment ministry anymore. It's a "whole-of-government" mission, pulling in finance, agriculture, construction, and transportation. The goal is to weave the value of a thriving ecosystem into the very DNA of every policy and every part of the economy.

Teaching the Calculators to Value a Sunrise

A revolution needs a treasury. Recognizing that public funds, however vast, could never be enough, China is now embarking on its most creative act: rewiring its financial nervous system to serve nature.

On the world stage, it has put its money where its mouth is. President Xi's pledge of CNY 1.5 billion (about $211 million) launched the Kunming Biodiversity Fund, a pool of money designed to help developing nations protect their own natural treasures. It's a clear signal of China's ambition to be a leader in global "South-South cooperation."

But the real magic is happening at home. China is exploring a concept so radical it could change how we measure success itself: Gross Ecosystem Product (GEP). For a century, we have worshipped at the altar of Gross Domestic Product (GDP), a metric that can see the value of a chopped-down forest sold as lumber but is blind to the value of that same forest left standing—the clean air it creates, the water it filters, the floods it prevents.

GEP is the antidote. It's an accounting system that puts a dollar value on the work nature does for free. How much is a wetland worth? GEP calculates the value of its storm protection. How much is a clean river worth? GEP measures the economic benefit of its water supply. Cities like Shenzhen are already piloting this system, and the results are being used to judge the performance of government officials. For the first time, a bureaucrat's career might depend not just on how much their economy grew, but on whether the birds returned to the river. It's an attempt to make the invisible visible, to finally balance the books between humanity and the planet.

China's journey is far from over. The scale of the environmental damage was immense, and the path to healing is fraught with challenges. But the direction has been chosen. China is building a new Great Wall—not of stone to keep out armies, but of redlines, regulations, and green finance, designed to protect the living systems that are its ultimate source of life. It is the most audacious act of planetary stewardship ever attempted, and its outcome will affect us all.

Australia - Climate Change & Invasive Species

The Silence and the Fever: Australia's Living Heart in Crisis

To know Australia is to feel a heartbeat in the red dust, to hear a story in the rustle of ancient leaves. This is a continent of deep time, a land that remembers the footfalls of megafauna and the first whispers of humanity. Its wildness is its soul, a tapestry woven with creatures spun from evolutionary fantasy—kangaroos that bound, platypuses that bewilder, and koalas that cling to a shrinking world. Australia is a global library of life, one of the precious few "megadiverse" nations. Yet, it is also a land of ghosts. It is the world's quiet champion of extinction, a place where the silence in the bush grows louder with each passing year. This is the story of a continent at the crossroads, where the weight of its history and the heat of its future are pressing down at once.

The Scars Upon the Skin: Wounds Old and New

The story of Australia's undoing is written in scars across its landscape, some old and faded, others fresh and weeping.

The Tyranny of the Straight Line: The deepest wound comes from a simple, brutal geometry: the fence line, the road, the paddock. Since colonization, the continent's subtle, flowing ecosystems have been carved and flayed by a relentless campaign of clearing. For agriculture, for cities, the ancient architecture of woodlands and grasslands was seen not as a living system, but as an obstacle. This isn't just history on a page; it is the headline reality in places like Queensland, a global deforestation hotspot. Each bulldozer blade tears more than trees from the soil. It shreds the web of life, stranding wildlife in shrinking "islands" of green. In these desperate archipelagos of survival, a koala finds its food source gone; a tiny marsupial faces the terror of an open field. The soil itself, stripped bare and vulnerable, bleeds into the rivers, its life-giving power lost to the wind.

The Uninvited Guests: Australia's isolation was its genius and its vulnerability. It was a self-contained world, and its creatures evolved without the cunning predators and aggressive competitors of elsewhere. Then the gates were thrown open. The newcomers—invasive species—have become a biological plague. The feral cat is not a pet; it is a shadow slipping through the undergrowth, a ruthlessly efficient killer responsible for billions of native lives each year. Its paws are stained with the final moments of species now lost forever. The red fox is a flicker of russet death for ground-dwelling birds and small mammals that never learned to fear such a hunter. And the invasion is not just of tooth and claw. Invasive "weeds" like gamba grass and lantana are a smothering green cancer, choking out native flora, poisoning the soil, and turning gentle bushfires into explosive infernos. This is a war fought metre by metre in every corner of the nation, a costly, heartbreaking battle to keep the native world from being completely overwritten.

The Great Fever: Draped over these chronic wounds is a new, acute crisis: the planetary fever of climate change. Australia is burning. The Black Summer of 2019-20 was not a bushfire season; it was a scream from a continent in agony. An area the size of a large country was incinerated. Three billion animals—a number so vast it numbs the mind—were killed or left homeless. We watched it on our screens, but many lived it in their lungs, as smoke turned day to apocalyptic twilight and ash settled like snow on suburban cars. For creatures like the Kangaroo Island dunnart, it was a near-total apocalypse, a trial by fire they may never recover from.

Out at sea, the fever takes a different form. The Great Barrier Reef, a living structure so vast it can be seen from space, is being boiled in a warming ocean. The phenomenon of coral bleaching is a gentle term for a gruesome process: the coral starves, turns a ghostly white, and dies. We are witnessing the real-time transformation of a vibrant underwater metropolis into a silent, crumbling boneyard. This is more than the loss of a tourism icon; it is the hollowing out of an entire aquatic universe, a tragedy visible in heartbreaking shades of white and grey.

A Shield Made of Paper: The Politics of Paralysis

Faced with a crisis of this magnitude, Australia has fashioned a shield made of paper. On the global stage, it whispers promises like the "30x30" target—a pledge to protect 30% of its land and sea. It's a noble ambition, but a line on a map is not a fortress. Many of our protected areas are "paper parks," starved of funds and staff, unable to fight off the invaders at their gates or the changing climate overhead. Protection without power is just a postcode for extinction.

The core of the problem lies in a collective failure of legal and political courage. The nation's premier environmental law, the EPBC Act, has been declared a broken, toothless guardian by one official review after another. A landmark 2020 report called for a complete overhaul, for legally binding national standards and an independent environmental watchdog—a "cop on the beat." The government nodded, agreed, and then progress slowed to a crawl, caught in the familiar crosscurrents of political cycles and industry lobbying. This failure to forge a strong legal shield means that, day by day, decisions are made that chip away at what's left, a death by a thousand approved cuts.

The Echo in Our Hearts: When the Land Aches, We Ache Too

This is not a story about trees and animals alone. It is about us. The unraveling of the natural world is the unraveling of our own prosperity, identity, and spirit.

The farmer scanning a cloudless sky, the tourism operator on the reef watching their vibrant world fade to white—their livelihoods are not separate from the environment; they are born from it. The natural services we take for granted—pollination for our crops, clean water in our dams, stable soil on our farms—are fraying, and the economic cost is mounting.

But the deeper cost is one of the soul. For First Nations peoples, who have tended this land for over 60,000 years, this is not degradation; it is desecration. The health of Country is the health of the people; every lost species, every bulldozed sacred site, is a severing of a spiritual connection that holds history, culture, and identity. For all Australians, the "bush" is woven into our sense of self. It is the sound of a kookaburra at dawn, the scent of eucalyptus after rain, the shared backdrop to our lives. The loss of this world creates a quiet anxiety, a low-grade hum of dread about the future we are creating. It is the grief of seeing the familiar become foreign, the living become memory.

Australia is not dying; it is being killed, by a thousand decisions, by a collective inertia. Yet, the continent's ancient heartbeat, though faint, endures. To save it requires more than new policies; it requires a new relationship. It means relearning the land's language, valuing its survival as our own, and finding the courage to become ancestors our descendants will be proud of. The silence in the bush is a warning. The fever in the ocean is a final call. The land is waiting.

Australia, we need to talk.

This isn't an intervention. It's a eulogy for things not yet gone, a love letter to a ghost that still breathes. It's an attempt to find words for the deep, aching grief of watching something magnificent come undone. You are a continent of impossible beauty, a living library of evolution, but your shelves are burning and the tide is seeping under the door.

This is the story of your beautiful, terrible paradox: a continent of fire and flood, of life and loss, all at once.

The Quiet in the Trees

The official reports read like a doctor's grim diagnosis, full of numbers that try to quantify a broken heart. They say over 2,000 of your children, your unique species, are now on a list, waiting for a final verdict. But a number can't capture the ghost-like flight of the Greater Glider, a creature made of fluff and moonlight, now silently vanishing as its cathedral of ancient gums is felled. A statistic doesn't convey the tragedy of a tiny, jewel-bright frog, whose entire universe is a single, cloud-wreathed mountaintop that is slowly becoming too warm to hold life.

The latest report offered a flicker of false hope—a season of good rain that painted your deserts in a blush of green. But we know this is the brief rally of a fever patient, a fleeting moment of clarity that doesn't change the underlying sickness. The diagnosis remains the same: a systemic failure driven by a climate in revolt, habitats erased with brutal efficiency, and invaders who thrive in the chaos.

It's not just the individual lives, either. It's the entire song. More than 100 of your ecological communities—the intricate orchestras of soil, fungi, flora, and fauna—are now listed as threatened. These are entire worlds, symphonies of co-existence perfected over millennia, from the sun-dappled woodlands of the east to the fragile alpine gardens of the Snowy Mountains. We are watching a library, assembled over eons, being erased in the space of our own breath.

A Fever Dream in Turquoise

Nowhere is your fever more apparent than in the turquoise dreamscape of the Great Barrier Reef. It is your soul rendered visible, a sprawling, living metropolis that could be seen from the heavens. Today, that metropolis is in the grip of a relentless heat.

Imagine holding your breath. Now imagine holding it until your lungs burn. This is what we've asked of your reef. As our world warms, your waters have absorbed the heat, triggering marine heatwaves that have become a near-annual torment. For the coral, this heat is poison. It forces them to expel the vibrant algae that are their lifeblood, their color, their energy. This is bleaching. It is the reef turning pale with starvation, a vibrant city turning into a bone-white ghost town.

The latest bleaching event was the second in two consecutive summers. There is no time to recover. It's like asking a marathon runner, already staggering from exhaustion, to immediately run another. And while the reef fights this fever, we are making its very bones brittle. The carbon dioxide we pour into the air dissolves into the ocean, turning it more acidic. This ocean osteoporosis makes it harder for corals to build their skeletons, weakening the very foundation of their world. The reef is being burned, starved, and dissolved, all at once. We are not just watching it die; we are watching it being unmade.

The Memory of Smoke

The scar of the "Black Summer" fires is seared into this nation's soul. It was a holocaust of trees and animals, a great smoke that choked cities and circled the globe. We speak of the three billion animals killed or displaced, a number so vast it becomes abstract. But it wasn't abstract. It was the silence where three billion heartbeats used to be. It was the smell of ash that lingered for months.

Today, the wounds are still weeping. More than half of all the species added to the threatened list since that summer were pushed there by the flames. The fires burned so hot in places that they sterilized the soil, ensuring nothing would grow back. For creatures like the tiny Kangaroo Island dunnart, the inferno consumed nearly their entire world, leaving a few singed survivors on a blackened rock.

This is your future in a warming world. The seasons of fire are getting longer, angrier, and more frequent. The ancient rhythm of fire and renewal has become a ritual of self-immolation. We are watching forests that have danced with fire for an eternity be consumed by an inferno they can no longer survive.

The Wound We Inflict Ourselves

And then there is the wound we inflict directly, the one we can't blame on a distant, warming globe. Australia, a first-world nation, is a global leader in deforestation. While the world watches the Amazon, we are quietly razing our own natural heritage.

Travel to Queensland and you can see it: a brutal efficiency that tears the heart out of the land. Two bulldozers, linked by a massive iron chain, drag a wall of steel across the earth. In their wake, a world is annihilated. Ancient trees, koala habitats, the homes of countless birds and insects—all reduced to a tangle of splinters, to be burned and replaced with grass for cattle. A staggering 88% of this clearing is for beef pasture. The steak on the dinner plate is a final, silent testament to a forest that no longer stands, a koala that no longer breathes.

This single act of destruction connects every part of your unraveling. It is the primary cause of extinction. It creates the tinder for the next megafire. It releases immense stores of carbon, fueling the fever in your oceans. And when the rains come, the cleared soil bleeds into the rivers, a torrent of mud and chemicals that flows out to the sea, choking the last breath from the corals of the Great Barrier Reef.

This is the state of our union, Australia. Your reef is connected to your forests, your forests are connected to your climate, and the survival of your unique and wondrous children is connected to us. You are a continent of breathtaking beauty, caught in a feedback loop of our own making. And one has to ask, when the song of the frogs fades, when the gliders no longer drift through the dark, and when the reef's colors are just a memory—what will be left of you? What will be left of us?

Of Dust and Dreams: Australia's Heartbeat at a Crossroads

An Inheritance of Sun and Silence

To know Australia is to feel it in your bones. It's the weight of the sun on your shoulders, the dry whisper of wind through ghost gums, the scent of eucalyptus after rain. This is not just a country or a continent; it is an inheritance, a living library of time itself, gifted to the present. For generations, we have built our modern lives, our ambitions, and our astonishing prosperity upon this ancient foundation. We've pulled wealth from its soil, drawn inspiration from its horizons, and sold its wild beauty to the world. We built a nation on the promise that this land was limitless.

But a deep and unsettling tremor now runs through that promise. We are discovering the truth that Indigenous cultures have always known: the land does not belong to us; we belong to it. A quiet, decades-long war, waged by our own hands and for our own comfort, is reaching its endgame. The complex, life-giving systems that power our economy and colour our identity are beginning to fray. We stand at a crossroads, forced to confront a terrifying question: what happens when the inheritance runs out, and we are left with only the dust of our dreams?

The Hunger of the Land

Listen. Beneath the hum of the city, there is another sound. It is the groan of hydraulics, the splintering of ancient wood, the deep shudder of the earth as another giant falls. This is the sound of our hunger. More than any other force, it is our desire to eat that is unmaking the wild continent. Agriculture, the proud backbone of the nation, has become the engine of its undoing.

At the heart of this destruction is a single, powerful appetite: our taste for beef. To feed it, we are clearing land on a scale that defies imagination, especially in Queensland, a place now infamous as a global deforestation frontier. The statistics are clinical, but the reality is a gaping wound. An area the size of the Sydney Opera House—a monument to our culture—is erased for pasture every four minutes. This isn't just logging; it's annihilation. It is the methodical erasure of entire worlds—the homes of bark-brown owls and shy wallabies, the tangle of roots that holds the soil, the canopy that breathes for us. With every cleared paddock, we tear a hole in the sky, releasing carbon that fuels a hotter, fiercer climate.

While the cattle stations carve the deepest scars, they are not alone. The story is echoed in the expansion of sheep paddocks, in the logging of irreplaceable old-growth forests for woodchips, in the gaping pits of open-cut mines that promise wealth but leave behind a poisoned sterility. It's in the slow, creeping sprawl of our suburbs, paving over bushland one cul-de-sac at a time. The logic is always the same: a patch of bush is seen as idle, unproductive, waiting to be "improved." This is the heartbreakingly simple math that is pushing our national icon, the koala, toward a silent disappearance in the east, its world literally bulldozed from under it. We are trading the irreplaceable for the consumable, and we are calling it progress.

The Uninvited

If the bulldozer represents the brute-force assault on our environment, an even more sinister threat has slipped past our defenses. It is a quieter, more patient invasion, an unravelling from within. Australia, the island ark, evolved in splendid isolation. Its creatures are unique, often gentle, and wholly unprepared for the cast of characters we brought with us in our ships. Now, these uninvited guests are waging a war of their own.

It is a war fought in the dead of night. The feral cat, a velvet-pawed assassin, and the European fox, a cunning hunter, have become a phantom menace in the bush. They move through the landscape like ghosts, leaving a trail of death that has already driven unique mammals like the desert rat-kangaroo into the darkness of extinction. For dozens of other creatures, from the rabbit-eared bilby to the painted numbat, these predators are a relentless, existential threat.

By day, the destruction continues. Hordes of feral rabbits, a plague unleashed from a handful of ancestors, strip the land bare, turning vibrant plains into dustbowls. The toxic cane toad, a biological blunder of epic proportions, marches its poisonous way across the north, killing any native predator foolish enough to try it for a meal. Feral pigs, goats, and horses churn wetlands into mud, foul sacred waterholes, and trample delicate soils.

And then there are the weeds. An insidious green tide of invasive plants that choke out the natives, transforming intricate mosaics of flora into dull, homogenous blankets of thorns and scrub. For Indigenous communities, this is more than an ecological loss; it is a cultural catastrophe. It is the loss of medicine plants, of bush foods, of the living landmarks that hold story and meaning. It is the slow erasure of a library thousands of years in the making, as the land itself becomes a stranger. The cost of fighting this "forever war"—a constant, losing battle of baiting, fencing, and spraying—is a tax paid in billions of dollars and immeasurable grief.

The Reckoning

The price of this broken relationship with the land is no longer an abstraction. The bill, long deferred, is now being slipped under our doors, in our mailboxes, in the smoke-hazed summer sky. The reckoning is here.

Our economy, which we imagined as a fortress of steel and data, is revealing its foundations of soil and water. The $88 billion agricultural sector, which drove the clearing, now watches its precious topsoil blow away on the wind and its river systems choke with silt. The tourism operator on the Great Barrier Reef, once a custodian of wonder, now sells trips to a ghostly, bleached graveyard, his livelihood as fragile as the coral itself.

A landmark study warned that if we continue on this path, the damage will carve US$20 billion from our economy every year by 2050. It's a loss that stings twice as hard as the global average, a sign of our unique vulnerability. The study revealed a terrifying truth: almost all of that loss comes from the death of our coastal ecosystems. We have been systematically dismantling the continent's natural armor—the reefs, mangroves, and seagrass beds that absorb the ocean's fury. Now, as the seas rise and storms intensify, we find ourselves a nation huddled by the sea, having sold its own shield for short-term gain.

The costs are no longer projections. They are the soaring insurance premiums after another "unprecedented" flood. They are the billions in taxpayer dollars spent cleaning up after fires made monstrous by a sick and fragmented landscape. They are the rising price of food, the scarcity of water, the gnawing anxiety of a future that feels less secure. The small profits reaped by clearing the land are being dwarfed by the colossal, ever-expanding cost of living on a continent we have wounded.

We are left with a choice that is about far more than economics. It is a choice about our identity. Are we a people who live on this land, or a people who live with it? Are we consumers, or are we custodians? The land is asking. The silence where the wild things used to be is waiting for our answer.

Australia's Choice: A Fraying Ark in a Rising Tide

Australia is more than a country; it's a living library of evolution, a continent of ancient dreams and fragile life. But this keeper of wonders is standing on a knife's edge. Its story today is one of profound conflict: a battle between soaring ambitions to save nature and the stubborn gravity of a broken system. It's a tale of whispered promises, starving budgets, and the quiet resilience of those who are rewriting the rules from the ground up.

The Paper Shield: A Law That Watches the World Burn

The nation's chief guardian is supposed to be a law: the Environment Protection and Biodiversity Conservation (EPBC) Act. Forged in 1999, it was meant to be a fortress, a legal stronghold for Australia's most treasured natural icons. In reality, it has become a ghost.

This isn't hyperbole; it's the cold, hard conclusion of two separate, exhaustive government-commissioned reviews. The most recent, led by Professor Graeme Samuel, didn't just find cracks in the system—it declared the entire foundation rotten. The Act, his report concluded, is a relic, fundamentally "unfit" for the modern crisis. It is not preventing destruction; it is presiding over it. It has become a masterful piece of legal theater, allowing the slow, methodical dismantling of the very heritage it was sworn to protect.

The Act's fatal flaw is its nearsightedness. It examines the world through a keyhole, judging each new mine or housing estate in isolation. It fails to see the web of life, the cumulative toll of each cut. This "death by a thousand cuts" has bled entire ecosystems dry. Astonishingly, the vast majority of land clearing in Australia—the primary killer of wildlife—never even comes before this national "guardian." It's a shield made of paper, offering the illusion of safety while the bulldozers roll on.

A new government swept to power on a wave of hope, promising to finally heed the warnings and build a new legal ark. They pledged strong, binding National Environmental Standards—a clear red line that could not be crossed. But the political winds shifted. That bold promise has withered on the vine, stalled in the corridors of power. And so, the broken law remains, the ghost in the machine, overseeing the decline, one approved project at a time.

A Grand Promise on a Shrinking Map

While its laws fail, Australia has made a breathtaking promise to the world. As part of the global "30x30" ambition, it has pledged to protect 30% of its land and 30% of its oceans by 2030. It is a promise of sanctuary, a race to build lifeboats before the storm hits.

At sea, the numbers look spectacular. Australia has designated a massive 45% of its ocean territory as Marine Protected Areas (MPAs). A victory, it seems. But look closer, and the mirage shimmers. The gold standard of protection—"no-take" sanctuaries where nature is left entirely to itself—covers only 18% of the marine estate. Much of the rest is open to various forms of fishing and exploitation. It's like declaring a library a quiet zone but still allowing shouting in most of the rooms.

On land, the challenge is a stark, physical reality. By mid-2025, about 24% of the continent is protected. To meet the 2030 deadline, Australia must find and safeguard another 40 million hectares. That is an area the size of Germany that needs to be brought into the conservation fold in the blink of an eye.

Where is the hope? It lies not in a government office in Canberra, but out on the country itself, in a revolutionary, world-leading movement: Indigenous Protected Areas (IPAs). These are not just parks; they are affirmations of a sacred relationship. Here, First Nations communities are the architects of conservation, weaving millennia of traditional knowledge with the tools of modern science. It is a partnership that heals both land and people.

IPAs are the beating heart of Australia's conservation estate, now accounting for more than half of its entire National Reserve System. They create jobs, revitalize culture, and prove that the original custodians of the land are its most effective guardians. As the clock ticks down to 2030, expanding the IPA program isn't just one option; it's the most profound and promising path forward.

The Price of Life and the Gospel of Markets

For all this ambition, Australia's conservation efforts are dying of thirst. The federal government's annual biodiversity spending has hovered for years at a paltry A$400-500 million. That's less than 0.05% of GDP—the fiscal equivalent of sending a firefighter to a blazing forest with a garden hose.

Experts, including Professor Samuel, estimate the real cost of turning the tide—of healing landscapes and pulling species back from the brink—is closer to A$7.3 billion a year. The gap between what's needed and what's spent isn't a crack; it's a chasm. The 2025 budget offered a temporary sip of water, but future projections show a swift return to drought.

Starved for cash, the government is turning to a new gospel: the market. The plan is to create a "Nature Repair Market," where corporations can buy "biodiversity certificates" from landholders who restore ecosystems. The vision is seductive: a new river of private capital flowing into conservation, funding the work of healing.

But it's a high-wire act. Can you really put a price tag on a wetland's soul? How do you create a "unit" of biodiversity without it becoming a token for corporate greenwashing? The integrity of this market is everything. If it works, it could spark a revolution in how we fund nature. If it fails, it will be a dangerous illusion, a distraction while the real-world ledger of extinction continues to grow.

Australia is a nation holding two futures in its hands. One is a future of managed decline, of paper shields and hollow promises. The other is a future of courageous investment, of true partnership with its First Peoples, and of a repaired relationship with the ancient continent it calls home. The choice is being made right now.

Spain - Desertification & Agricultural Intensification

Beneath the postcard image of sun-drenched fiestas and flamenco dancers, another Spain is emerging—a Spain of cracked earth and thirsty riverbeds. The country's vibrant, passionate soul is becoming dangerously parched. It's a land standing at a precipice, forced to confront a future it has long ignored, one where its most celebrated assets—the sun and the soil—have become instruments of a slow, creeping demise.

The Silent Fever of the Land

To feel the pulse of modern Spain is to feel a landscape in the grip of a fever. More than two-thirds of the country is naturally dry, a geography of sun-baked plains and stark mountains that has always demanded respect. But now, that dryness is deepening into a chronic illness. This isn't the dramatic march of Saharan dunes; it's a quieter, more insidious death. It is desertification: the soil's life force bleeding away, its structure crumbling to dust, its promise of fertility turning to a barren memory. A staggering 74% of Spain is now caught in this struggle, a patient showing severe symptoms.

Climate change acts not as a gentle nudge but as a violent accelerant, pouring gasoline on an already smoldering fire. The heat is no longer just a feature of summer but a year-round siege. Rain, when it comes, is often a brief, furious deluge that scars the hardened ground rather than a gentle shower that soothes it. Droughts have ceased to be cyclical events and have become a near-permanent state of being, especially in the sun-scorched southern heartlands of Andalusia, Murcia, and Valencia. Here, the land is not just changing;

it is being fundamentally rewritten into a harsher, more desolate script.

This environmental siege is triggering a silent, mass exodus. Spain, a veritable Noah's Ark of European biodiversity, is watching its inhabitants vanish. In what sounds like a dispatch from a war zone, recent surveys show that up to 60% of Spain's native species are now threatened. The crisis screams loudest from its waterways. The nation's rivers and lakes, the very arteries of the ecosystem, are failing. An astonishing 55% of its freshwater fish and 31% of its amphibians are on a trajectory toward extinction, their fate tied to the water that is simply no longer there.

A recent study in Catalonia put a number on this heartbreak: a 28% collapse in wildlife populations in just two decades. The decline wasn't random; it was a targeted hit. In the very places where humanity's hand is heaviest—the vast, industrialized farms and the diverted, dammed-up rivers—the living world has been decimated, with populations plummeting by 40%.

A Thirst We Created

This is not a natural disaster; it is a human-made tragedy, born from a collective decision to prioritize explosive growth over sustainable life. The two main architects of this crisis are an agricultural system built on a fantasy of infinite water and a tourism model that sells a slice of paradise at an unsustainable price.

Spanish agriculture became a miracle built on borrowed time—and borrowed water. Encouraged by policies that rewarded quantity over all else, Spain transformed its driest regions into Europe's vegetable garden. The "Miracle of Almería" is a sea of plastic greenhouses shimmering under the sun, a testament to human ingenuity. But it is also a crime scene. This agricultural machine is ferociously thirsty, drinking up 80% of the nation's water. It has relentlessly pumped life from ancient underground aquifers, leaving them depleted, sometimes poisoned with saltwater, and unable to feed the wetlands and rivers that depend on them. Iconic sanctuaries like Doñana National Park are being slowly starved to death to quench the thirst of strawberry and avocado fields.

At the same time, Spain's coastline was remade into a playground for the world. But the concrete march of hotels, golf courses, and swimming pools has been a brutal conquest. It has paved over fragile habitats, erasing three-quarters of the Mediterranean's coastal sand dunes. The idyllic vacation, with its long showers and sparkling pools, creates a staggering water debt, as a single tourist can consume many times more water than a local resident—a demand that peaks in the driest, hottest months.

A Cure We Are Afraid to Take

Spain is not without a plan. Guided by the EU, it has created one of Europe's largest networks of protected areas, Natura 2000, a paper shield covering over a quarter of the country. National strategies to fight desertification and protect biodiversity are written with noble intent. But these plans often feel like well-meaning doctor's orders ignored by a patient in denial. The protected status of a park like Doñana means little when the water that gives it life is being siphoned off just outside its borders.

The ultimate reckoning must be with water. For generations, Spain's answer to scarcity was engineering: build a bigger dam, a longer canal, and move water from the wet north to the thirsty south. This approach, which fueled the economic booms of the 20th century, has hit a wall. There is simply not enough water left to move, and the ecological price of these manipulations has become devastatingly clear.

The only way forward is a radical paradigm shift. The nation must pivot from a frantic search for more supply to a disciplined management of its demand. This means asking the unthinkable: Is it sane to grow water-guzzling crops in a near-desert? It means making water pricing reflect its true, life-giving value, not treating it as a cheap, inexhaustible input.

The most potent solutions may lie not in new technology but in old wisdom. Healing the land itself—replanting forests, restoring wetlands, letting rivers flow freely—can rebuild the natural systems that capture and store water. A healthy ecosystem is the best reservoir.

Spain is being asked to choose its identity. Will it cling to the illusion of a green and verdant paradise built on unsustainable consumption, or will it find a new pride and a new prosperity in making peace with its arid soul? The challenge is no longer technical; it is a test of courage and imagination. It requires a new national story, one that redefines wealth not by what can be extracted from the land, but by what can be sustained. It is the wisdom to understand that in a dry land, water is not a commodity, but a sacred trust.

Spain is eating itself alive.

It began as a feast. A glorious banquet fueled by sun, soil, and ambition. The world came to its shores, drawn by the scent of orange blossoms and the promise of endless summer. Its fields, a powerhouse for a continent, churned out a bounty of olives, fruits, and wine. For a time, it seemed the prosperity would never end.

But the feast has become a fever. The nation, in its relentless pursuit of more, has turned its fork upon itself. The very lifeblood of the land—its water, its soil, its wildness—is being devoured to satisfy a fleeting, insatiable appetite. This is not a story about climate change statistics or policy debates. This is a story about a land's soul being sold for short-term profit, a story etched into the parched riverbeds of Andalusia and the worried faces of its farmers.

Imagine standing in a place that was once an olive grove, a true grove, where gnarled, ancient trees stood as silver-green patriarchs. Between them grew wild herbs and flowers, a universe for bees and beetles. Larks nested in the tall grass, their songs stitching the sky to the earth. Now, look again. The old grove is gone, bulldozed for a new kind of olive farm: a "super-high-density" grid of stick-thin trees, marching in perfect, sterile rows to the horizon. There are no flowers, no birdsong, no life between the lines. There is only the hum of irrigation pumps and the chemical gleam of pesticides. This is not farming; it is a factory laid upon the corpse of a field. We gained efficiency, but we lost the symphony.

This relentless drive for yield has created a nation with a raging, unquenchable thirst. Agriculture, the prodigal son, consumes nearly 80% of Spain's water, demanding it in the driest places, at the driest times. We drill into the earth, deep into the ancient aquifers—the land's sacred savings accounts, filled drop by drop over millennia—and drain them in a single generation to grow strawberries in the desert.

Nowhere is this tragedy more poignant than in the Doñana wetlands, one of Europe's last great wildernesses. It is a place of mythic beauty, a cathedral of water and wings where millions of birds rest on their epic journeys. But the cathedral is crumbling. Its altars—the shimmering lagoons—are turning to cracked mud, choked by the thirst of the surrounding berry farms. It is a jewel being dissolved by the very people sworn to protect it, a slow-motion sacrilege driven by the global demand for a sweet, red fruit out of season.

This thirst now creeps from the fields into the cities. In the villages of Catalonia, taps run dry. Grandmothers teach their grandchildren the forgotten art of conserving every drop, a lesson their own parents thought was a relic of a poorer past. Ships bearing drinking water—a once-unthinkable sight—now dock in bustling ports. A quiet war is being waged over what remains, pitting the farmer against the city dweller, the golf course against the family home. Who has the greater right to the last drops of a dwindling inheritance?

And then there is the coast, loved so fiercely it has been smothered. We saw the sun-kissed beaches and, instead of preserving them, we entombed them in concrete. A relentless tide of hotels, holiday villas, and apartment blocks has replaced the dunes and wetlands that once protected the shore. We have built a paradise that consumes itself. A single tourist, enjoying a poolside holiday, can use three times more water than the local resident whose well is running dry. We have created a bizarre paradox: an economy that relies on selling a vision of natural beauty while systematically paving it over.

This is the human cost of a broken bond with the earth. It is the hollowed-out villages of "España vaciada," where the young have fled because the soil can no longer support a family. It is the farmer watching his ancestral land, once rich and dark, turn to pale dust that blows away in the wind. It is the loss of a shared identity, a connection to a landscape that has shaped the Spanish character for centuries.

Spain's story is a profound and urgent warning. It shows that an economy severed from its ecological roots is not an economy at all, but a pyramid scheme destined to collapse. The country stands at a precipice, forced to confront the ghosts of its choices. The path forward demands more than new policies; it requires a revolution of the heart. It requires remembering that the true wealth of a nation is not in its quarterly profits, but in the health of its lands, the purity of its waters, and the wild, untamed beauty that, once lost, can never be reclaimed. The feast is over. Now, the bill has come due.

A World in the Balance: Convergent Crises, Systemic Risks, and the Post-2025 Imperative

The Thread of Desperation

This thread is pulled not by greed, but by an empty stomach. It is the story of survival, the most intimate and tragic kind of destruction. Imagine a father in Madagascar. He stands at the edge of one of the planet's last, most magical forests, a living museum where 90% of the creatures exist nowhere else on Earth. He is not a villain. He is a provider. He holds a machete not out of malice, but because his family has not eaten. The forest is his only pantry, his only field. The ancient practice of tavy—slashing and burning a small patch of irreplaceable jungle to grow rice—is a choice made when there are no other choices. As the smoke curls into the sky and another piece of the tapestry turns to ash, a lemur's world shrinks. For him, the long-term cost of extinction is a debt he cannot afford to consider when his children's hunger is a bill that is due tonight.

Now, drift across the ocean to the Coral Triangle, the underwater Amazon. Picture a fisherman in his small boat, floating above a reef that was once a kaleidoscope of life. Foreign factory ships, like silent leviathans, have already plundered the deep, leaving him with empty nets. His desperation leads him to a terrible calculus. A stick of dynamite, a splash of cyanide. The explosion is a thunderclap of finality, shattering a century of coral growth in a heartbeat. For a fleeting moment, the water is filled with stunned, easy-to-catch fish. It is a bounty, but it is also a funeral. He has bombed the very cradle of life he depends on, poisoning the nursery that would have birthed the fish of tomorrow. This is the death spiral of survival: a world of beauty and sustenance dismantled piece by piece, by the very hands that need it most.

The Thread of Progress, Unspooled

The second thread is pulled by a far more insidious force, one that whispers of efficiency and progress. This is the story of intensification, of wringing every last drop of production from lands we have already tamed. It looks like success, but it is the most elegant and devastating illusion.

Fly over the American Midwest. From above, it is a breathtaking quilt of green and gold, a testament to human ingenuity. But descend closer, and you feel the silence. This is not a landscape; it is a factory floor. The native prairies, once a riotous sea of grasses and wildflowers buzzing with life, have been sterilized into a monoculture of corn and soy. To keep this fragile system from collapsing under the weight of pests, it is drenched in a chemical cocktail of pesticides and herbicides. The ghosts of butterflies now drift through the sterile rows. The bees, our essential partners in creation, are dying in droves, their small bodies riddled with poisons like neonicotinoids.

And the illusion ripples outward. The nitrogen fertilizers that create this unnatural abundance do not stay put. They bleed into the water, a slow poison seeping through streams and rivers, all the way to the Gulf of Mexico. There, they fuel a monstrous algal bloom that suffocates the ocean itself, creating a "dead zone" the size of a small country—a wound in the sea where nothing can live. We have become brilliant at maximizing the yield of a single stalk of corn, but we have done so by bankrupting the ecological capital—the clean water, the living soil, the pollinators—that makes growth possible in the first place. We have engineered a perfect harvest in a dying world.

When the Tapestry Tears

When enough threads are pulled, the fabric doesn't just fray; it tears. The local acts of destruction—a burned patch of forest, a poisoned reef, a silent field—are now converging into a cascade of global, systemic collapse. This is no longer an "environmental issue." It is the precipice upon which our economies, our stability, and our health now teeter.

The quiet, thankless work of pollinators puts between \$235 and \$577 billion worth of food on our tables each year. As they vanish, our food system begins to tremble. The world's largest insurance companies, the ultimate pragmatists, are sounding the alarm. They see that in destroying the mangroves and reefs that buffer our coasts, we are making ourselves uninsurably vulnerable to the next storm. They are not acting out of love for nature, but out of a calculated fear of bankruptcy.

This instability bleeds into our societies. When the land turns to dust and the wells run dry, people must move. The farmer from the Sahel, the fisher from a bleached reef—they become migrants, fueling social and political crises far from their homes. And as we tear down the walls between our world and the wild, we invite ancient plagues into our lives. COVID-19 was a brutal lesson in how a disturbed ecosystem can unleash a pathogen that brings the entire globe to its knees. Scientists warn us that there are over a million undiscovered viruses lurking in the wild. Our destruction is not just clearing forests; it is rolling the dice with a global pandemic.

And then there is the final terror: the tipping point. The Amazon rainforest, the great lung of our planet, may be approaching a point where it can no longer generate its own rain. If it crosses that threshold, it will not gently decline; it will collapse, transforming into a dry savanna and releasing a "carbon bomb" that will make climate change unstoppable. This is the moment the tapestry ceases to be a garment and becomes a shroud.

Mending the Fabric

In the face of this, we gather. We sign pacts like the Kunming-Montreal Global Biodiversity Framework, ambitious and beautiful in their intent. Yet we live in a state of profound paradox. With one hand, we pledge to protect nature; with the other, we sign trade deals that demand more of the very commodities that fuel its destruction. We ask Brazil to save the Amazon while the world screams for its beef and soy.

But to pull a thread is not the only power we have. We can also learn to weave. New threads of hope are emerging. A revolution in finance is beginning to demand that money tell the truth about its impact on the living world. Debt-for-nature swaps are offering nations a way out of financial crisis by investing in their greatest asset: their wilderness. In courtrooms across the globe, a quiet, radical idea is taking hold: the Rights of Nature. This movement grants a forest or a river the legal right to exist and defend itself, shifting our relationship with the world from one of ownership to one of kinship.

The evidence is in. The story is clear. The unravelling is not something happening to us; it is a story being written by us, by the choices we have made and the systems we have built. The next few years are our final, fleeting chance to change the ending. Success will not be found in another treaty, but in a revolution of the human heart—in rewiring our economies and our laws to recognize the simple, staggering truth that we are not separate from the tapestry. We are the tapestry. We are the weavers now. The threads are in our hands.

Our War Against Ourselves: The Two Faces of Nature's Destruction

Listen. Can you hear it? It is the sound of the world growing quiet. It is the silence where a forest hummed, the stillness where a reef once danced with color. We talk of saving the planet, but the story we tell is incomplete. We cast corporations as the sole villains, their smokestacks and bulldozers the obvious weapons in a war against nature. But that is only one front in a conflict waged on two sides, a pincer movement closing in on the wild heart of our world.

This is a war with two faces. One is etched with desperation, its battlefields carved out by the empty-handed, the hungry, the people for whom tomorrow is a luxury they cannot afford. The other face is comfortable, complacent, its campaigns waged not with an axe, but with a credit card and a demand for relentless convenience. It is a war fueled by both the empty stomach and the full shopping cart. And if we are ever to declare peace, we must first look both of our combatants in the eye.

The Battle for the Next Meal

Imagine the forest is the only grocery store, the only pharmacy, the only power plant you have ever known. This is not a metaphor; it is the daily calculus of survival for millions. In the world's most vibrant, fragile ecosystems, the greatest destructive force isn't malice, but necessity. It is the cumulative weight of countless arms, each reaching for the branch that will cook a meal, the patch of land that will grow enough food to quell a child's hunger.

In the Democratic Republic of Congo, a nation that breathes for the planet, the great rainforests are not falling to industrial chainsaws. They are being nibbled to death. Ninety percent of the loss is the work of families, ghosts in the global economic machine, clearing just enough land for a season's worth of cassava, burning just enough wood to see through the night. When this quiet, desperate act is multiplied by a hundred million people, the sound becomes a roar, and one of Earth's vital lungs begins to suffocate.

Fly to Madagascar, and you can see the island bleeding. From space, its rivers run like open wounds, red with the soil of a dying landscape. This is the home of creatures spun from fantasy—lemurs with startled eyes, chameleons of impossible color. But it is also home to one of the world's poorest peoples. Their ancestral farming practice, tavy, a careful dance of slash-and-burn, was once sustainable. But squeezed by a growing population, the dance has become a frantic stomp. Fallow periods shrink, hillsides are stripped bare, and the land gives up its ghost. To survive, farmers must push deeper, chasing a fertility that is always receding. In this tragic spiral, the Malagasy people, in the simple act of feeding their families, are unintentionally erasing a chapter from the book of life.

And then there is conflict—the gasoline on the fire. In the eastern DRC, the gun and the charcoal kiln have become partners. Rebel groups fund their wars by clear-cutting protected forests, while the people displaced by the violence flee into these same sanctuaries, forced to consume the very nature that was meant to be saved. In Colombia, a celebrated peace deal had a perverse effect. As FARC rebels left their jungle strongholds, a power vacuum sucked in a swarm of land-grabbers who understood that the fastest way to claim territory was to burn the forest to the ground. Peace, tragically, became a green light for destruction.

In these places, conservation is not about putting up a fence. It is about building a bridge out of poverty. It is about peace. It is about offering a choice other than the one between your child's future and the forest's.

The Quiet Violence of Convenience

On the other side of the world, the weapons change. The destruction is quieter, more sterile. It is the violence of the spreadsheet, the hum of the mega-farm, the relentless pursuit of an efficiency that sees life as a bug in the system. Here, in the lands of plenty, nature is not being consumed for survival; it is being suffocated by the sheer intensity of our lifestyle.

Look at the American heartland. It is a miracle of production, a sea of corn and soy stretching to a ruler-straight horizon. But this green expanse is a green desert. Watered with chemical cocktails and managed by GPS, this system pumps out cheap food at an ecological cost we are only now beginning to calculate. The soil turns to dust, the chemicals bleed into the rivers, creating vast "dead zones" in the ocean where nothing can breathe. This agricultural machine, a marvel of human ingenuity, is a primary threat to more than 86% of the nation's endangered species.

Across Europe, the story is the same. In Spain, the demand for year-round strawberries and tomatoes has scrubbed the landscape clean of its "unproductive" parts—the hedgerows, the wildflowers, the messy bits of life that once gave shelter to birds and bees. The countryside has fallen silent, with farmland bird populations crashing by more than a quarter in a few short decades. On the coasts, a tidal wave of tourism has paved over ancient dunes with concrete resorts and golf courses, sucking precious water from an already thirsty land.

Our cities and roads are the shears that snip the fabric of the wild into useless scraps. A highway slices through a migration path that has existed for millennia, turning a contiguous wilderness into a series of isolated, inbred biological islands. And hitching a ride on our global trade routes are the biological mercenaries—invasive species. In the U.S., they are a silent plague. In Australia, a continent of evolutionary oddities, they have been an apocalypse. The European fox and the feral cat, introduced by settlers, arrived as super-predators in a land with no defenses. They are the chief architects of Australia's heartbreaking mammal extinction rate, ghosting through a bush that was not made for them, leaving only silence in their wake.

The Fire Raining on Both Fronts

Looming over both the desperate farmer and the comfortable consumer is the great threat multiplier: climate change. It is not a separate war; it is the atmosphere itself turning hostile, amplifying every existing pressure to the breaking point.

It is the force that turned Australia's seasonal fires into the "Black Summer" inferno that incinerated three billion animals. It is the warmth that bleaches the life from the Great Barrier Reef. It is the drought that turns the Amazon, the lungs of our planet, into a tinderbox, pushing it toward a catastrophic tipping point. And in Madagascar, it is the hand that withholds the rains, creating the world's first climate-famine and tightening the very noose of poverty that drives the chainsaws.

The climate crisis and the biodiversity crisis are not two problems. They are the same problem, seen from the ground and from the sky. A planet stripped of its forests, wetlands, and reefs has no shield against the coming storm. A planet in the grip of that storm will see its last wild places burn, flood, and wither.

This is the war we are all waging. A war against the ecosystems that give us air to breathe and water to drink. A war fought with an empty stomach in one hemisphere and a thoughtless purchase in the other. It is a war on nature, which means it has always been a war on ourselves. And it is the only war in human history where our unconditional surrender—our laying down of arms and finally choosing to nurture rather than to conquer—is our only path to victory.

Our Planet's Overdrawn Account: The Bill Has Come Due

For generations, we've been telling ourselves a story. A comforting bedtime story for a global economy that couldn't sleep soundly if it faced the truth. The story was simple: we could take from the Earth without consequence. Its forests were just timber, its rivers just plumbing, its oceans just a pantry. We treated our planet like an infinite trust fund, drawing down its capital while ignoring the frantic, red-inked warnings on our statements.

That story is now over. The final chapter is being written not in books, but in bleached coral, in silent forests, and in the exhausted faces of people fleeing lands that can no longer support them. This is not a distant, academic tragedy about polar bears. This is a story about us. The intricate web of life is not a backdrop to our lives; it is the very loom upon which our societies, our health, and our economies are woven. And as we sever thread after thread, the entire tapestry is beginning to unravel.

The Auditor at the Door: Nature's Economic Reckoning

Imagine nature as our silent business partner. For millennia, this partner has provided essential services for free. Insects pollinating crops worth billions. Forests stabilizing the weather that allows our farms to thrive. Reefs and mangroves acting as a free sea wall, protecting our cities from storms. We never put these services on our balance sheets. We saw them as entitlements, not assets. In the greatest accounting error in human history, we valued them at zero.

Now, the auditor is at the door, and the numbers are terrifying. The World Bank estimates that the collapse of these "free" services could vaporize $2.7 trillion from the global economy every single year by 2030. This isn't a forecast; it's an invoice for services we took for granted.

Look at Brazil. The breathtaking Amazon rainforest, the lungs of our planet, provides the rainfall that underpins much of South America's agriculture. Yet, the short-term profits from clear-cutting it for cattle and soy are celebrated. An analysis revealed the devastating irony: the potential long-term economic losses from the Amazon's collapse—a staggering $317 billion annually—are seven times greater than the profits gained from destroying it. It is economic madness, like burning down your house to sell the scrap metal.

This pattern of self-harm is a global pandemic. Malaysia, a nation built on its natural wealth, could see its GDP shrink by 6% by 2030 as its ecosystems fail. Australia faces a potential $20 billion annual loss by 2050 if it continues its "business as usual" approach.

The risk has now moved from the farm field to the trading floor. Central bankers, the most cautious people in the world, are sounding the alarm. In Brazil, 46% of all corporate loans are held by businesses that depend directly on a healthy environment. In Malaysia, that number is 54%. As nature's services crumble, so does the collateral. The risk of a cascade of loan defaults is no longer theoretical. It is a clear and present danger to global financial stability. Protecting a watershed is now a vital act of risk management.

The Parable of the Bat: A Chain of Devastation

If these vast numbers are hard to grasp, consider a single, devastating chain of events. In the United States, bats are a farmer's best friend, a silent, flying pest-control service that works for free every night. Then, a fungus causing White-Nose Syndrome began to decimate their populations.

Between 2006 and 2017, as the bats vanished, farmers were forced to fight the resurgent insects themselves. Their use of chemical pesticides skyrocketed by 31%. The direct cost in lost crops and chemicals was a jaw-dropping $26.9 billion.

But the bill didn't end there. In the counties most affected, the increased wash of pesticides into the environment had a horrific side effect: infant mortality rose by 7.9%. The calculated social cost of this human tragedy was another $12.4 billion.

Think about that. A microscopic fungus on a bat's nose leads directly to billions in economic damage and the loss of human life. This is the brutal calculus of a broken ecosystem. We are learning, in the hardest way possible, that when we liquidate our planet's natural assets, the price is paid not just in dollars, but in lives.

When the Foundation Cracks: The Human Cost of Collapse

The consequences of this global unraveling are not distributed fairly. They are a wrecking ball aimed at the world's most vulnerable. The cost is measured in conflict, displacement, and disease.

In Colombia, the phrase "environmental conflict" is not an abstraction. It is a reality where community leaders and Indigenous peoples are murdered for defending the land and water that gives them life. In the Democratic Republic of Congo (DRC), the fight for control of mines that feed our hunger for cobalt and minerals fuels unspeakable violence and human rights abuses.

In southern Madagascar, a land scorched by drought and desertification, a famine born of climate change has forced countless people to abandon their ancestral homes in a desperate search for survival. In Nigeria, the Sahara Desert marches south like an invading army, pushing farmers from their lands into overcrowded cities, igniting social fires. This isn't just a distant problem. When a nation like Indonesia decides to move its entire capital from Jakarta—a city sinking, choking on pollution, and besieged by rising seas—it is an admission of environmental defeat on a monumental scale.

The link to our own bodies is just as stark. In Nigeria's Niger Delta, decades of oil spills have left a legacy of poison, causing a 60% drop in household food security and a 24% spike in childhood malnutrition. In the DRC, mining pollution has been tied to an epidemic of birth defects.

And as we tear down forests, we are kicking open the door to new pandemics, bringing ourselves into closer contact with wildlife and the pathogens they carry—the very process that likely unleashed horrors like HIV and Ebola upon the world.

On the Edge of Forever: Planetary Tipping Points

The most chilling revelation is that we may be pushing some of the Earth's most vital systems toward a cliff edge—a point of no return.

The Amazon is our most urgent and frightening example. Scientists warn that if 20-25% of the rainforest is destroyed, it could trigger its own death spiral. The forest, which creates its own rain, would lose that ability and transform into a dry, barren savanna. We are perilously close, with 14-17% already gone and vast areas weakened. In 2023, the world watched in horror as over 300 river dolphins boiled to death in a record-hot Amazonian lake—a shocking cry of distress from the heart of the forest. Still, the assault continues. A plan to pave the BR-319 highway, a dagger aimed at the forest's untouched core, could be the final push that sends it over the edge.

Half a world away, Australia's Great Barrier Reef is being cooked alive by climate change. It has now suffered six mass bleaching events since 2016. It is losing its color, its complexity, its very soul. It is becoming a ghost of its former self, a simpler, algae-choked graveyard.

The collapse of these ecosystems would not be a local event. It would trigger a global shockwave. The death of the Amazon would destabilize weather across a continent, threatening the breadbaskets of Argentina and Paraguay. The loss of the Great Barrier Reef would expose Australia's coast to the full fury of the ocean. Protecting these places is not a national choice; it is a global imperative.

The Architect's Dilemma: Forging Our Path Forward

So what do we do? We are caught in a great paradox. Governments make bold promises on the world stage, then fly home to sign policies that fund the very destruction they vowed to prevent.

Brazil, the future host of a major climate summit, funnels billions in agricultural loans to properties illegally clearing the forest.

Indonesia has plans to clear millions of hectares for plantations while talking of conservation.

Courts in Brazil and Colombia have become environmental heroes, declaring the government's failure to act "unconstitutional" and even granting the Amazon rainforest legal rights of its own.

In Ecuador, citizens bypassed politicians entirely, voting in a historic referendum to halt oil drilling in a precious national park.

The biggest obstacle is not a lack of solutions, but a lack of coherence. We celebrate a $450 million "debt-for-nature" swap to protect the Galápagos—a brilliant innovation. But that brilliant drop in the bucket is dwarfed by the tidal wave of money flowing in the opposite direction. Globally, we spend an obscene $542 billion every year on harmful subsidies that pay industries to pollute, deforest, and destroy.

This points to the most powerful strategy we have. While raising new money for nature is good, the single most impactful act would be to stop funding the destruction. Redirecting those hundreds of billions in harmful subsidies is the masterstroke. It would not only free up immense capital for healing, but it would turn off the main financial engine of our planet's demise.

Choosing Our Tipping Point

This is a story of two tipping points. We are hurtling toward the negative ones—the irreversible collapse of the systems that keep us alive. But we are also seeing flashes of positive tipping points—a courageous court ruling, a citizen uprising, a new financial model that values a living tree more than a dead one.

The path forward requires three commitments. First, we must recognize that rights and equity are not optional. The most successful conservation on Earth happens when the rights and wisdom of Indigenous Peoples and Local Communities are at the center. Second, we need strong laws and real enforcement. Voluntary corporate promises have failed. Third, and most crucially, we need policy coherence. The finance ministry and the environment ministry must work for the same team—Team Humanity.

We are standing at the precipice, caught between collapse and recovery. The bill for our long party has come due. How we choose to settle this account will define the future of life on Earth. It will be the single greatest story of our time, and we are all its authors. Let us choose to write a story of recovery, of courage, and of a world rebuilt on a foundation of respect for the home that gives us everything.

About Author

Azhar ul Haque Sario is a bestselling author and data scientist with a remarkable record of achievement. This Cambridge alumnus brings a wealth of knowledge to his work, holding an MBA, ACCA (Knowledge Level - FTMS College Malaysia and London College of Accountancy, London), BBA, and several Google certifications, including specializations in Google Data Analytics, Google Digital Marketing & E-commerce, and Google Project Management.

With ten years of business experience, Azhar combines practical expertise with his impressive academic background to craft insightful books. His prolific writing has resulted in an astounding 2810 published titles, earning him the record for the maximum Kindle editions and paperback books published by an individual author in one year, awarded by Asia Books of Records in 2024.

ORCID: https://orcid.org/0009-0004-8629-830X
Azhar.sario@hotmail.co.uk
https://www.linkedin.com/in/azharulhaquesario/